THE WILDERNESS MOVEMENT AND THE NATIONAL FORESTS

by

Dennis M. Roth

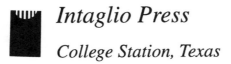

Intaglio Press

College Station, Texas

Library of Congress Cataloging in Publication
Data Roth, Dennis Morrow, 1943-
The Wilderness movement and the national Forest

1. Wilderness areas -- United States -- History
2. Forest reserves -- United States -- History
3. Forest policy -- United States -- History
4. United States. Forest Service -- History. I. Title
   QH76.R68 1995 333.78 '2' 0973

   ISBN 0-944091-05-9 (pbk.)

Photographic Credits: National Agricultural Library
USDA Forest Service

Manufactured in the United States

(Revised)
SECOND EDITION

Cover: Hancock Lake, Marble Mountain Wilderness Area
Klamath National Forest, California
Courtesy National Agricultural Library

Presidential Range, Dry River Wilderness Area
White Mountain National Forest, New Hampshire

CONTENTS

Acknowledgements

Foreword

Chapter I	The National Forests and the Campaign for Wilderness	1
Chapter II	The Modern Environmental Movement	11
Chapter III	Primitive Area Reviews and The Parker Decision	17
Chapter IV	The Lincoln-Scapegoat- The First De Facto Bill	25
Chapter V	RARE I and the 1975 Act on Eastern Areas	53
Chapter VI	The Endangered American Wilderness Act and RARE II	61
Chapter VII	Oregon Wilderness	79
Chapter VIII	An End, and a New Beginning	89

Appendices

    A. National Wilderness Preservation System 91

    B. Designated Wilderness Areas, Forest Service 92

Notes

ACKNOWLEDGEMENTS

This book originally was published in two parts: Chapters 1 and 8 appeared in the July 1984 issue of the *Journal of Forest History* (vol. 28, no. 3). Chapters 2 to 7 appeared as *The Wilderness Movement and the National Forests: 1964-80* (Washington, D.C.: Forest Service, FS 391-1984.)

The Second Edition revises the first edition published by Intaglio Press with the inclusion of materials from the author's *The Wilderness Movement and the National Forests: 1980-1984* (Washington, D.C.: Forest Service, FS 410, 1988). Tim Mahoney, the librarians of the Denver Public Library, Holly Jones, and Peter Coppleman were especially helpful. I received excellent cooperation from all the individuals and organizations contacted.

FOREWORD

In the early 1920's Forest Service employees created the first wilderness areas in the world. By the late 1930's conservation organizations began to assume leadership in the wilderness movement. For the last forty years no issues have affected the Forest Service more so than the preservation of wilderness areas and, more recently, the protection of "old growth" forests. Because of these issues, Congress and the public have become intimately involved in the daily activities of the agency. This is a history of how that came to be.

Dennis Roth

Chapter 1
The Campaign for Wilderness Legislation

The Wilderness Act, signed by President Lyndon B. Johnson on September 3, 1964, is a landmark in conservation history. The act provided statutory protection for more than 9 million acres of recreational wilderness and charged three federal agencies - the U.S. Forest Service, the National Park Service, and the Fish and Wildlife Service - with the enormous task of reviewing potential wilderness areas in their jurisdictions. Confrontations between preservationists and commercial interests commanded the lion's share of public attention during the dramatic eight-year congressional battle that resulted in the Wilderness Act. However, the federal agencies that were to assume responsibility for implementing the new national wilderness preservation system also played a crucial part in shaping the bill's history.

Long before the legislative battle for wilderness began, the Forest Service had designated more than seventy areas within the western national forests for special wilderness management. As the federal agency with the largest percentage of potential wilderness in its jurisdiction, the Forest Service played a key role in forging the federal response to pressure for wilderness management. Moreover, the agency carried a broad multiple-use mandate, and its growing commitment to wilderness was subject to strong and varied interest-group pressures. Unlike the Park Service and the Fish and Wildlife Service, whose priorities were closely linked to traditional nonutilitarian preservation concepts, the Forest Service was involved in profound and controversial reassessments of its forest management philosophy throughout the various stages of the wilderness movement. Thus, the history of the Wilderness Act is largely a history of national forest wilderness politics.

Until the early decades of the twentieth century, wilderness conditions had existed as a by-product of the movement to preserve scenic wonders in national parks and to reserve from the public domain some forests and rangeland so that watersheds and future supplies of timber could be protected. As a specific land classification, wilderness did not exist in its own right; rather, it was embedded in other land-management categories.[1] In the late nineteenth century national park advocates had shifted their interests from "natural wonders" to broader notions of preserving scenic and primeval landscapes, and passage of the National Park Act in 1916 was in part a tribute to the impulse to preserve scenic wilderness. Yet, the act was as much a mandate for development as it was for preservation. It called upon the newly-created Park Service to protect the natural integrity of the parks but at the same time make them accessible for use by the public.[2] This contradiction was latent, for automobiles were still a novelty in 1916 and park tourism was only a fraction of what it would become a few decades later. But the promotional zeal that would become so characteristic of national park administration as a result of the 1916 act curbed the service's role as a pioneer in wilderness management planning.

Ironically, it was the Forest Service which initiated experiments in federal wilderness management in the second decade of the twentieth century. In keeping with its concept of multiple-use planning, the Forest Service recognized recreational activities in its jurisdiction as early as 1905, although recreation ranked well below commercial use in the agency's priorities. In 1917 some 3 million recreationists visited the national forests, and in that year the Forest Service commissioned a landscape architect, Frank Waugh, to survey the forests for their recreational potential.

The shift from recreational to wilderness concerns came shortly thereafter; the Forest Service's growing attention to wilderness in the 1920's and 1930's was a response to a number of pressures from public and industrial interests and the competition with the Park Service for recreational budgets and land. But in large part the call for wilderness designation came from within the agency itself. In the first half of the twentieth century, Aldo Leopold, Arthur H. Carhart, and Robert Marshall designed and helped implement a wilderness policy for the Forest Service. Doing so, they created a defense of wilderness that would figure prominently in the creation of the National Wilderness Preservation System in 1964.

Aldo Leopold was a young forest supervisor on the Carson National Forest in New Mexico. An eastern-trained intellectual who loved to ride the backcountry and hunt, he had discussed the idea of setting aside wilderness areas with Elliott Barker, one of his rangers on the Carson, as early as 1913.[3] At the end of World War I, Leopold had become apprehensive about the expansion of Forest Service road systems into the backcountry. Wilderness, he felt, was the forge upon which the American national character had been created, and loss of wilderness regions deprived the country of a source of renewing this heritage. Leopold had also been inspired by the new science of ecology which considered wilderness an ideal laboratory for the study of natural processes.[4] In 1918 the Ecological Society of America was formed, and in 1920 Francis B. Summer, an eminent student of ecology and genetics, urged the society to back the setting aside of untouched areas. A year later, Leopold seconded this call in a *Journal of Forestry* article that suggested a wilderness of at least 500,000 acres for each of the eleven states west of the Great Plains.[5]

In May 1922, Leopold, now assistant district forester in Albuquerque, made an inspection trip into the headwaters of the Gila River. When he returned, he wrote a wilderness plan for the area that excluded roads and additional use permits, except for grazing. Only trails and telephone lines, to be used in case of forest fires, were to be permitted. Leopold's plan encountered opposition from some of his own colleagues in the district office who thought that development should take precedence over preservation. However, on June 3, 1924, District Forester Frank Pooler approved Leopold's concept. The Gila area was to be placed under a ten-year wilderness recreation policy; grazing and waterpower developments were not to be impeded, but roads were to be limited and efforts were to be made to acquire private inholdings through land exchanges. Pooler's action did not carry immediate national signficance because the Forest Service's Washington Office was not involved in the decision, but the 500,000-acre Gila Wilderness would serve as a precedent for wilderness designation in other areas in the coming decades.[6]

In 1924 Leopold was transferred from the Southwest to the Forest Products Laboratory in Madison, Wisconsin. He retired from the Forest Service four years later. During his career of nearly twenty years with the agency, Leopold formulated a sophisticated defense of wilderness and its values. Before his death in 1948 he would serve in the leadership of more than 100 professional societies and conservation organizations, help to train a generation of conservation and wildlife leaders at the University of Wisconsin, and produce *A Sand County Almanac*, a classic in American conservation literature. His impact upon the Forest Service's emerging wilderness management policy was profound.

Leopold looked at wilderness through the eyes of a hunter and budding ecologist. In 1919 the Forest Service commissioned the services of Arthur Carhart, a landscape architect who would emphasize the aesthetic and recreational benefits of a policy of nondevelopment. Carhart was assigned to the district office in Denver and was asked to survey the Trappers Lake area in Colorado's White River National Forest for summer homesites. While laying out his survey lines, he came upon two wealthy hunters, Paul J. Rainey and William McFadden, who convinced him that the area should remain wild and that houses and roads should be excluded from the vicinity of the lake. Carhart's training as a landscape architect and his affinity for unspoiled nature left him receptive to this argument. As a result, he developed a functional recreational plan to preserve the area's pristine conditions. Carhart convinced his superiors to halt plans to develop the region, marking, in the words of Donald Baldwin, "an unprecedented step in Forest Service history."[7]

Carhart followed his success at Trappers Lake with a memo to Leopold after a meeting between the two men in the Denver District Office. "The Forest Service," Carhart admitted, "is obligated to make the greatest return from the total Forests to the people. . .that is possible," but, he concluded,

> *There is. . .a great wealth of recreational facilities and scenic values within the Forests, which have not been so utilized and at the present time the Service is face to face with a question of big policies, big plans, and big utilization for these values and areas.*[8]

Carhart later surveyed the lake region of the Superior National Forest in Minnesota and repeated his recommendation that development be minimized and that the area be preserved as much as possible for primitive travel by canoe. On September 9, 1926, Secretary of Agriculture William H. Jardine signed a plan to protect the area, providing a wilderness policy decision at the highest departmental level. In 1964 the area was dedicated as the Boundary Waters Canoe Area.

Carhart left the Forest Service at the end of 1922, and Leopold's departure two years later for the Forest Products Laboratory left the Forest Service's evolving wilderness policy in the hands of others. The strongest impulse behind the movement continued to be growing pressure for recreational use of the national forests. Forest Service annual reports from 1921 to 1929 reflect an increasing awareness of the importance of recreation. Beginning in 1926, wilderness too was singled out for separate discussion. According to Chief William B. Greeley, the wilderness idea had "merit and deserves careful study, but its correlation with the other obligations and requirements of National Forest administration must be carefully worked out before definite steps are taken to give any areas a wilderness status." The following year Greeley wrote more positively about wilderness areas, announcing that the Forest Service planned "to withhold these areas against unnecessary road building" and to prohibit forms of commercial use that would impair their "wilderness character."[9] In 1926 Greeley and Assistant Forester for Lands Leon F. Kneipp ordered an inventory of all undeveloped national forest areas larger than 230,400 acres (ten townships). Three years later wilderness policy assumed national scope with the promulgation of the L-20 Regulations covering so-called national forest primitive areas. Management priorities were defined "to maintain primitive conditions of transportation, subsistence, habitation, and environment to the fullest degree compatible with their highest public use with a view to conserving the values of such areas for purposes of public education and recreation." A management plan was to be prepared which would show the conditions under which timber, forage, or water resources would be utilized, permanent improvements constructed, and special-use occupancy allowed.[10]

According to James P. Gilligan, whose 1953 doctoral dissertation was to play an important role in the development of the wilderness bill, these regulations were simply strong recommendations to Forest Service field personnel suggesting limitations on unplanned development in untouched areas. They did not categorically prohibit any form of development or use, and, consistent with the agency's decentralized style of administration, they gave a great deal of discretion to field personnel. To illustrate, Gilligan pointed out that by 1939 some 14 million acres of national forest land had been designated "primitive." Nevertheless, as late as 1953 logging, grazing, and roads had been totally excluded from only four primitive areas.[11]

Historians have suggested that the Forest Service had two primary reasons for recommending the L-20 Regulations. First, they were to discourage personnel from rushing into unnecessary development projects. Second, they were to stem the transfer of more land to Park Service units. Most of the national parks since 1916 had been carved out of national forest acreage.[12] Designating primitive areas showed that the Forest Service too was interested in preserving wild regions and that land transfers to the Park Service were unnecessary.

Some Forest Service officials have denied the importance of interagency rivalry in explaining the L-20 Regulations. Many areas, they suggest, were transferred to the Park Service despite primitive designation; L-20 proved to be no guarantee that the land would be retained by the Forest Service.[13] Nevertheless, it is possible that primitive designations forestalled creation of even more parks. For example, despite vigorous campaigning the Park Service never convinced Congress to convert portions of the Idaho national forests to park status, perhaps because the Forest Service had previously designated large sections of this area - now part of its Frank Church River-of-No-Return Wilderness - as primitive. The Forest Service's reluctance to place an acreage minimum on primitive areas also suggests an intent to discourage parks. Leopold had suggested 500,000 acres and Kneipp had asked for an inventory of areas exceeding 230,400 acres, but the L-20 Regulations permitted instead many smaller designations. Although this may have been an early recognition that wilderness can be protected in relatively small areas, the suspicion persists that it was an attempt to encompass any area that might have national park potential.[14] This interagency rivalry was clearly reflected in bureaucratic terminology. In the 1920's the Forest Service called its undeveloped areas "wilderness," while the Park Service called its backcountry "primitive." When the Forest Service published its L-20 Regulations covering "primitive areas," the Park Service switched to "wilderness."[15]

Preservation organizations, such as the Sierra Club and The Wilderness Society, were later able to take advantage of this rivalry. Given its legislative mandate to preserve scenic landscapes, the Park Service naturally received strong support from wilderness advocates, but the Forest Service too found it expedient to court this growing element in the public lands constituency. Moreover, the Forest Service could press its advantage by asserting that it would not clutter its primitive areas with recreational developments, which many wilderness advocates were finding disturbingly prevalent in the national parks by the 1930's.[16]

Few were more skilled at exploiting this situation than Robert Marshall. The son of a wealthy New York attorney, Marshall dreamed of wilderness exploration while growing up on Park Avenue and hiking in the Adirondacks. Afraid that he had been born too late to emulate his boyhood heroes, Lewis and Clark, he felt compelled to climb every peak and explore every wilderness possible. After a short stint in the Forest Service (1925-1928), he earned a Ph.D. in plant physiology and became chief forester for the Office (now Bureau) of Indian Affairs, where he established a system of Indian wilderness. From his position with the Department of the Interior, he exhorted the Forest Service to increase its primitive acreage. In 1933 he wrote the recreation section of *A National Plan for American Forestry* (the Copeland Report), thereby gaining broader public exposure for his wilderness ideas.

While working within the BIA, Marshall pressed his conservation-minded boss and friend, Secretary of the Interior Harold L. Ickes, for a stronger Park Service commitment to wilderness preservation. An astute and ambitious administrator, Ickes hoped to expand his domain by enlarging the national parks at the expense of the national forests. Thus, Ickes was disposed to listen to Marshall, and in 1939 he proposed legislative protection for park wilderness. Icke's plan died, however, primarily because the National Parks Association, the service's main support organization, feared that special protection for park wilderness would result in the degradation of heavily used areas of the parks.[17]

Marshall left the Interior Department in 1937, disappointed by what he saw as the Park Service's growing attachment to recreational development. He was recruited to head the Forest Service's Lands Division, which had recently been renamed Recreation and Lands. From this new position within the Forest Service, Marshall spoke out for stronger wilderness protection, focusing his advice particularly upon Chief Forester Ferdinand A. Silcox.

In 1939 the U Regulations, drafted by Marshall and approved by Silcox, supplanted L-20. The new regulations afforded much greater protection to wilderness areas, prohibiting timbering, road construction, and special-use permits for hotels, stores, resorts, summer homes, organization camps, and hunting and fishing camps. The use of motorboats or the landing of aircraft, except where such practices were well established or were needed in emergencies or for administration, were also prohibited. Livestock grazing was permitted, as it had been under L-20 because stockmen would have vigorously opposed any ban. Water storage projects were also allowed if they did not involve road construction, as were improvements necessary for fire protection, "subject to such restrictions as the Chief deems desirable." Forest Service wilderness areas, like the rest of the national forest system, remained subject to existing mining and leasing laws and to the possibility of reservoir and dam construction. The Forest Service could only insist under the U Regulations that such developments be undertaken with a minimum of damage to the wilderness environment.[18]

In many instances, primitive areas had been created without adequate surveys. Thus, under the U Regulations all primitive areas were to be re-evaluated before they could be reclassified by the secretary of agriculture as "wilderness areas" (100,000 acres or more), or by the chief of the Forest Service as "wild areas" (5,000 to 100,000 acres). In addition, the public was given opportunity to comment during a ninety-day period prior to reclassification. In the interim, primitive areas were to be given protection under the U Regulations.[19]

In his brief two and a half years with the Forest Service, Marshall had succeeded in strengthening its wilderness policies, but he became disenchanted by the slow pace of reclassification and the failure to enlarge the potential bank of wilderness areas beyond 14 million acres. Before his death in 1939, he was considering leaving the Forest Service in order to apply pressure from outside the government.[20]

Shortly after Marshall's death the Forest Service entered a new phase of timber management. Timber had replaced grazing as the service's largest source of income by the mid-1920's, but the annual cut was insignificant compared to the potential harvest; many roadless areas were preserved in their pristine state simply because there was no demand for their resources. Rapid expansion of the industry at the end of World War II and overcutting on many private forests compelled private timber owners to turn to the national forests to supply their needs; the Forest Service thus entered an era of intensive management and became involved in the market economy.

In addition to the clamor for wood-based commodities, the revived post-war economy brought more people to the national forests and increased the demand for recreational facilities. Despite the formulation and expansion of wilderness designations in the 1930's, preservationists entered the postwar era with a growing feeling of unease over the long-term security of the wilderness areas. Bob Marshall's death had deprived them of their last important bridge to the Forest Service, and in the face of rising recreational and commercial demands on the parks and forests, preservationists grew anxious about continued wilderness protection based on administrative decisions alone. What one administrator could do, another could undo. They especially feared policy changes by new Forest Service leaders, who seemed willing to interpret the concept of "wise use" flexibly to meet increased demands for timber in the postwar world. Moreover, preservationists distrusted the agency's decentralized organization, which worked admirably in allowing personnel to adjust to local social and economic conditions, but which, they felt, militated against the application of uniform standards for the protection of fragile wilderness resources. Finally, preservationists sensed a crucial difference in the "mental image" of wilderness held by foresters, who often looked upon environmental changes as subject to the recuperative powers of nature. Wilderness advocates abhorred such changes as a threat to the idea of purity that sustained their image of wilderness.[21] The next decade would see a campaign for congressional protection that would ensure the gains made at the departmental and agency levels.

All of the structural criticisms that the preservationists raised regarding wilderness management in the 1940's and early 1950's became focused on the issue of Forest Service primitive area reclassification. They had applauded the Forest Service when it placed all primitive areas under the U Regulations, but they began to express doubts over the process of reclassification. Partially because of the curtailment of many activities during World War II, primitive areas were being assessed at a slow rate. (Only 2 million acres had become wilderness by the late 1940's). In some cases, critics claimed, primitive areas reclassified as wilderness lost their timbered sections in lower elevations. The Forest Service could point out that the total primitive acreage had remained stable, but the preservationists countered that this had been done by substituting high-elevation, nontimbered, rocky terrain for forested sections, thus degrading the quality and variety of the wilderness areas.[22] One preservationist wag opined that the Forest Service had agreed to a wilderness policy of "no timber harvesting above timberline."

Perhaps the best-known example of the defects of reclassification took place on the Willamette National Forest in the mid-1950's. The Forest Service removed 53,000 acres of timbered land at lower elevations in the Three Sisters Primitive Area. This action aroused the ire of Oregon's junior senator, Richard L. Neuberger, who soon after became a cosponsor for the wilderness bill. Neuberger remained the bill's most ardent congressional advocate until his death in 1960.[23]

Wilderness advocates were aware that the Forest Service was under pressure from local communities to increase the amount of land available for timber harvesting. Ever since the creation of the national forests in the late nineteenth century, administrators had been criticized by westerners for "locking up" resources. Sensitive to these pressures, preservationists began to explore ways in which the Forest Service and Park Service could successfully resist development and give free rein to their best wilderness intentions. In addition, preservationists felt that they had to combat what they perceived as a deterministic attitude among foresters who felt that however worthwhile wilderness preservation might be, it must give way to economic development.[24]

Wilderness areas not covered by U Regulations were disappearing at a rapid rate. At least 35 million acres of "de facto" wilderness in the national forests had been developed between the Kneipp survey of 1926 and 1960.[25] Continued development seemed certain, but the preservations differed with the determinists in that they believed that many such de facto areas could be preserved by congressional action. This conviction later became the philosophical preamble of the Wilderness Act, which was to ensure that "an increasing population, accompanied by expanding settlement and growing mechanization, does not occupy and modify, all areas within the United States and its possessions, leaving no lands designated for preservation and protection in their natural condition. . . ."[26]

In 1946 Howard C. Zahniser of The Wilderness Society drew together these and other concerns in his criticism of an American Forestry Association conference report presented at Higgins Lake, Michigan. To Zahniser, the conference recommendations appeared only to rationalize -- "with . . . up-to-date terminology" -- increased commercial use of America's forests. If this assumption proved correct, he intimated, he and his fellow conservationists would be compelled to "charge our expectations up to disillusionment and enter a prolonged period of fighting for wilderness preservation with our guard up whenever 'good forestry' is mentioned . . . ." Zahniser responded with a call for wilderness zoning for certain areas of "primitive America".[27] He was probably thinking of some form of legislation to compel zoning, although he did not say as much to the foresters. It was indeed such a statutory mandate that Aldo Leopold had suggested when Zahniser was hired as executive secretary of The Wilderness Society and editor of its publication, *Living Wilderness*.[28]

Zahniser was born in Franklin, Pennsylvania, in 1906, the son of a Free Methodist minister to whom Zahniser attributed the evangelical fervor of his own crusade for wilderness. He joined the federal government in the U.S. Biological Service in 1930, and from 1931 to 1945 served as a writer, editor, and broadcaster for several conservation agencies. In 1945 he accepted a position as executive secretary (later executive director) of The Wilderness Society. With Aldo Leopold's death in 1948, Zahniser became the nation's foremost wilderness advocate and assumed most of the burden of articulating a philosophy and program of wilderness protection until his own death at fifty-eight, a few months before passage of the Wilderness Act.[29]

Zahniser's enthusiasm corresponded with the concerns of a growing number of people who were discovering the attractions of wilderness after the Second World War. Not a backcountry adventurer like Marshall and Leopold, the bald and bespectacled Zahniser looked more like a stereotypical librarian. A bookstore was always his first stop when visiting a new city. A modest income left Zahniser a more appropriate role model for the average wilderness lover than the wealthy, backpacking "elitists" who figured prominently in the demonology created by wilderness opponents. Although he derived great inspiration from nature, as shown by his moving account of a trip to the Cloud Peak Primitive Area in Wyoming in 1947, he seldom found the opportunity to get away from work. When he suffered a heart attack in 1952, he blamed himself for not taking his own advice on the therapeutic value of wilderness.[30]

In a 1947 exchange of letters with a Berkeley forestry professor, Frederick S. Baker, Zahniser expressed several of the philosophical themes that he would repeat over the next seventeen years. He accepted neither Baker's fatalistic assumption that wilderness would inevitably disappear nor his solipsism that wilderness is "within us" and need not depend on a pristine physical reality. For Zahniser, wilderness was indeed "within us", but the mental image itself was "dependent on the perpetuation of wilderness," just as, he might have said, Plato's idea of Beauty was grounded in the contemplation of beautiful forms. Zahniser likened wilderness areas to art museums, a simile he used often in the following years. Like the museums, wilderness areas contained national treasures; even people not destined to see them firsthand could appreciate the effort to protect them.[31]

To Baker's argument that wilderness enthusiasts were only interested in self-gratification and escape, Zahniser countered that wilderness was also morally uplifting.

> *Love of solitude, eagerness for adventure, and indulgence in romantic experience are, as you point out, the most common motives for 'fleeing to wilderness' for recreation. Once there, however, many I believe, experience a better understanding of themselves in relation to the whole community of life on earth and rather earnestly compare their civilized living with natural realities-to the improvement of their civilization.*[32]

As Stephen Fox has suggested, this moral view of wilderness, hardly evident before World War II, may have been accentuated by the implications of living in a nuclear age.[33] Nature must uplift man lest he destroy it along with himself.

Zahniser's letters to Baker, written to elicit the forestry professor's views, reveal two personality traits that would serve his cause admirably. Zahniser was patient but determined, and he was pragmatic. Although he adamantly defended the moral value of wilderness, he recognized that the concept meant different things to different people. Flexibility and persistence enabled him to follow the wilderness bill doggedly for seven years through sixty-six rewrites and many compromises. Even then, he wore himself out and, like Marshall and Leopold, died at a relatively young age.[34]

One of Zahniser's earliest tasks after joining The Wilderness Society was helping President Benton MacKaye draft a bill to create "wilderness belts" throughout the United States. According to the MacKaye plan, a nonprofit citizens' group, supervised by a committee consisting of representatives from the Park Service, Forest Service, and Fish and Wildlife Service, was to use congressional appropriations to acquire land that would be reserved as wilderness and protected from all forms of development.[35] This plan did not include the federal lands that were of most concern to preservationists, but it marked a beginning as the first specific proposal for a legislated national wilderness system. Although the bill was never introduced to Congress, it gave Zahniser valuable practical experience in the mechanics of drafting a bill. By 1947 The Wilderness Society had spoken publicly in favor of legislative protection for federal wilderness, but Zahniser withheld specific proposals until he had mustered a consensus among other major conservation organizations.

In 1949 Zahniser had his first opportunity to bring the issue to the attention of conservation groups. Through the good offices of Carl D. Shoemaker of the National Wildlife Federation, Zahniser persuaded Congressman Raymond H. Burke, chairman of the Subcommittee on Wildlife Resources of the House Committee on Merchant Marine and Fisheries, to commission the Library of Congress to complete a study of America's wilderness needs. The study was assigned to C. Frank Keyser, who sent questionnaires to federal and state agencies and conservation organizations.

The Forest Service's response pointed to mining and reservoir construction as the largest threats to its wilderness areas, but it recommended further study before any legislative action was taken. The more sanguine Park Service said it managed parks as units and kept recreational developments to a minimum; it suggested only that more wilderness land be transferred to its jurisdiction. The Fish and Wildlife Service, which managed the nation's wildlife refuges under authorities less clearly defined than those governing the Park Service and the Forest Service, favored some kind of national wilderness policy. After analyzing the responses, including a forty-six page report from Zahniser, Keyser concluded that majority opinion favored national legislation for wilderness. The report was published as a House Committee print, and a limited number of copies were given to The Wilderness Society for distribution to those interested.

Keyser's report set the stage for Zahniser's first detailed proposal for federal wilderness legislation. His plan, offered at the Sierra Club's Second Biennial Wilderness Conference, included several points he would later incorporate into the initial wilderness bill. Congress, he suggested, should establish a national wilderness system; it should define the proper uses of areas within the system and prohibit incompatible uses; it should identify appropriate areas and list areas qualified for possible inclusion later; it should specify that additions could be made by executive order, but that removal of areas from the system could be effected only by Congress; it should make clear that no changes in jurisdiction over wilderness areas would be involved - that the agency administering a designated wilderness area would simply be charged with responsibility for preserving its wilderness character; and finally it should authorize a commission to recommend to Congress any necessary adjustments in the program. Zahniser's choice of the term "system" emphasized his preference for comprehensive wilderness legislation, in contrast to the piecemeal "foot-in-the-door" strategy advocated by other perservationists.

The unveiling of Zahniser's plan coincided with the beginnings of a controversy that would fundamentally alter the wilderness movement. Before 1950, threats to wilderness had come in the form of the Park Service's recreational developments and the Forest Service's gradual declassification of land at lower elevations in primitive areas - issues that bothered preservationists but drew little attention from the general public. Threats of a more dramatic nature emerged in the late 1940s with a series of proposals from the Bureau of Reclamation and Army Corps of Engineers for dams in several western national park units. Water users proposed, among other things, a dam at Echo Park in Dinosaur National Monument, to be a part of the ambitious Upper Colorado River Storage Project. For preservationists, the Echo Park proposal rekindled memories of the bitter Hetch Hetchy defeat and the loss of that beautiful valley in Yosemite National Park some forty years earlier.

Dinosaur National Monument, located in northwestern Colorado and eastern Utah, is an isolated and starkly beautiful region known primarily for its valuable paleontological remains. It had been proclaimed as a national monument in 1915 and considerably enlarged by President Franklin Roosevelt in 1938. Preservationists were outraged by the dam proposal and launched what would become a five-year campaign to thwart it. By the time the proposal had been deleted from the storage project in 1955, it had provoked more public opposition than any issue in recent conservation history. Preservationist organizations and the public, skillfully mobilized by advocates such as David R. Brower of the Sierra Club and Zahniser, responded not only to the Echo Park proposal but to what they perceived as a threat to the entire national park system.

The Echo Park controversy clearly demonstrated that in the absence of an aroused public opinion the national park system was ultimately unable to protect wilderness values, and that a new, more comprehensive system was needed. The campaign had great educational value as well; although focused on a national park unit as little known as many of the Forest Service primitive areas, the Echo Park campaign had succeeded in rallying massive public support for wilderness values. The very anonymity of Echo Park had special significance; had the dispute involved a better-known park unit, it may not have highlighted the need for comprehensive wilderness legislation.

Conservation organizations came away from the Echo Park controversy greatly encouraged. The struggle had united the movement more than ever; the need for wilderness preservation was gaining recognition among the general public, and Zahniser had developed new contacts in Congress and new lobbying skills to be employed on behalf of the wilderness bill.

In May 1955 Zahniser began an effort that political scientist James Sundquist said "followed the classic pattern for developing support for a bill." [36] In a speech to an American Planning and Civic Association meeting in Washington, D.C., Zahniser presented the philosophical arguments for wilderness preservation and repeated the specific proposals he had made in 1951. Senator Hubert H. Humphrey, who had been involved in the fight to preserve Echo Park, inserted the speech in the Congressional Record. The Wilderness Society mailed reprints of the speech to its members and those in other conservation organizations under the franks of cooperative legislators. Humphrey was so impressed with the strength of the wilderness sentiment that he asked Zahniser to develop a bill. Although he may have wished to wait a little longer, Zahniser began work on his first draft in February 1956. Within weeks the document was circulating among the leaders in the conservation movement.

Zahniser and his collaborators expressed three main concerns. They wanted a clear, unambiguous bill, free of loopholes. Second, they wanted to maintain the coalition that had formed earlier to protect Dinosaur National Monument. Thus, Zahniser was scrupulous in circulating early drafts to the major conservation organizations. The third and most difficult task was minimizing the opposition. Tactical considerations aside, the bill's drafters hoped to provide clear statutory authority for the maintenance of wilderness areas, to remove Forest Service authority for declassifying or decreasing the size of wilderness-type areas, to protect national forest wilderness against mining and water projects (both of which were authorized by Congress), and to require designation of wilderness zones in units of the national park system, the federal wildlife refuge and range system, and within Indian reservations.[37]

In order to gain support from federal land-managing agencies, Zahniser specified that existing jurisdictions would be respected and that the bill would not supersede the purposes for which the land was being administered, except to require that its wilderness character be preserved. Moreover, as Senator Humphrey later pointed out, it was not to be a "reform" measure but a bill that merely encouraged and sanctioned the good practices of the country's land-managing agencies.

Zahniser attempted to blunt opposition from commodity interests by assuring them that existing uses would be respected. This, however, was of little comfort to stockmen or operators of motorboats or airplanes because their uses of wilderness were considered "nonconforming" and were to be terminated in a manner "equitable" to them. The bill contained language that could be used to enforce condemnations if the agencies were unable to reach voluntary agreements with the nonconforming users.

The bill's supporters gave much thought to two provisions that would be hotly contested in the ensuing national debate and that would be primarily responsible for delaying passage more than eight years. These concerned mining and the manner in which Forest Service primitive areas could be added to the wilderness system. (The wilderness and wild areas were to go immediately into the system.) Zahniser's first draft had prohibited mining and prospecting except on claims already established. In a later draft he liberalized this section by allowing the President to open wilderness areas to prospecting and mining if necessary for the common defense and security of the nation. (National park wilderness, most of which had never been open to mining, remained under the categorical ban.) This was done to disarm critics who charged that the wilderness system endangered national security. However, on the advice of Lyle F. Watts, former Forest Service chief, this provision was deleted and Zahniser reverted to his original position. Needless to say, mining interests emerged as the most vigorous opponents of the wilderness bill.

In his first draft, Zahniser gave congressional protection to the Forest Service U Regulations as they applied to wilderness and wild areas. The Forest Service was given until January 1, 1965, to reclassify all of its primitive areas. Previous practice was altered only in that congressional authority was necessary to modify or eliminate wilderness areas. In a later draft this procedure was dropped in favor of a "legislative veto." Under this scheme, proposals from the executive branch that would modify, eliminate, or create a wilderness area could be stopped by a majority vote of either house. The procedure was consistent with the preservationists' view of the federal agencies: they usually did the right thing but had to be stopped when they went astray. It had the effect of increasing the initiative of the executive branch and placing Congress in the role of a somewhat passive reviewer, but as far as the preservationists were concerned, it combined the best of both worlds. On one hand it provided fast and professional administrative action, unencumbered by legislative logjams or a committee chairman's power to delay legislation. On the other hand, Congress retained the authority to react quickly to public dissatisfaction should the executive branch attempt an unpopular decision. Opponents of the wilderness bill would argue that the legislative veto was inappropriate in the case of public lands and that it surrendered congressional authority granted by the Constitution. They successfully pushed for affirmative action by both houses of Congress.

Underlying this constitutional issue was a debate over how much land should go into wilderness. In 1963 Washington Senator Henry M. Jackson succinctly stated the matter to one of his constituents.

In my opinion, either way of doing the Congressional review gives the right of either House of Congress to block the recommendations of an area as wilderness. The real fight is between the conservationists' desire for a mechanism which will force Congress to act to keep an area out of the wilderness system, and the effort of the opponents to require Congressional action before an area gets in. The proponents want to prevent wilderness proposals from dying because Congress fails to get around to them. The opponents want to capitalize on delays and oversights to keep areas out. It is a struggle for tactical advantage . . .[38]

At the end of February 1956, copies of the draft bill were informally given to the Park Service and the Forest Service. Conrad L. Wirth, director of the Park Service, replied that such a bill was not necessary and might even endanger national park wilderness areas by lumping them together with those of other agencies. The Forest Service opposed the provision establishing a wilderness commission (later changed to "council") to monitor the land-managing agencies. (The council was finally deleted in 1961.) Administrators also feared that other special interests would seek similar legislative guarantees for their uses of the national forests. In the late 1940's and early 1950's stockmen had attempted precisely that. Their push for virtual proprietory rights over national forest rangelands had been blocked by 1956, but the nearly successful effort was still fresh in the minds of Forest Service leaders. Ironically, Arthur Carhart, a citizen of Colorado where the "range war" had been most intense, at first opposed the wilderness bill for that reason.[39]

In early March the bill was introduced to a wider circle in New Orleans at the Twenty-first Wildlife Conference, the major annual forum for conservationists. One month later it received its full public unveiling at the Northwest Wilderness Conference, sponsored by the Federation of Western Outdoor Clubs, which promptly wrote Hubert Humphrey urging its passage. Impressed by the federation's enthusiastic support, Richard Neuberger asked to cosponsor the bill, which was introduced on the Senate floor on June 7. By then support and opposition for S. 4013 had already begun to coalesce. Because it was late in the legislative season and the presidential election was only five months off, Humphrey introduced the measure only as a study bill. The full legislative process did not begin until February 1957, when Humphrey reintroduced the bill as S. 1176.

The bill applied to several areas of possible wilderness classification. Along with immediate inclusion of national forest wilderness, and wild areas, the bill called for study and consideration of regions in the national forest primitive areas and in the national parks, the national wildlife and game regions, and Indian reservations.

National park wilderness had, with some exceptions, always been withdrawn from nonrecreational commercial development, and had the bill been concerned only with the parks it would have encountered much less opposition, even though the the Park Service did not fully accept the idea of statutory protection for its wilderness areas until 1960. Wilderness in national wildlife and game refuges was somewhat more contentious because the bill contained language that some construed as allowing the president to create additional refuges. Under one version of the bill, these would become part of the wilderness system if Congress did not veto the proposals. Commodity groups saw the specter of an unlimited expansion of wilderness areas created by presidential fiat and ratified by congressional inaction. In 1961 this procedure was limited to those refuges already in existence, which partially mollified the bills' opponents.[40] Indian wilderness areas were deleted in 1961 when it became apparent that Indian tribes and the Bureau of Indian Affairs were strongly opposed, even though the bill gave tribal councils the right to veto the designation of any Indian land as wilderness. Earlier, Bob Marshall had created an Indian wilderness system without the tribes' consent, and understandably they did not want further federal interference with their wilderness lands.[41]

The legislative history of the bill can be summarized as a process of drawing together support for the wilderness system and chipping away at the opposition through a series of compromises aimed primarily at commercial users of the national forest lands. Regulations regarding grazing best illustrate this process. Grazing is the oldest and best-established commercial use of national forest areas. Until the 1920's grazing fees were the largest source of income from all national forest system lands. Stockmen were a potent political force in the West and exerted their power whenever the Forest Service threatened to raise grazing fees or cut back on overgrazing. Under these circumstances the Forest Service had allowed controlled grazing in wilderness areas under the L-20 and U Regulations.

In his 1949 report to the Library of Congress, Zahniser had called grazing a "nonconforming" use which should be terminated. In 1956 he incorporated this language into the wilderness bill with the provision that terminations be made "equitably." In 1957 the bill was amended to state that grazing "may" continue under such "restrictions" as the secretary of agriculture deemed desirable. Finally, in 1963 "may" was changed to "shall" and "restrictions" to "regulations."[42]

But even with grazing recognized as a valid use of national forest wilderness, stockmen feared that they would be denied the money-saving advantages of roads and motorized equipment. Moreover, they, like other commodity groups, had an amorphous fear that wilderness legislation somehow contained a hidden gremlin or that it could be used to expand the wilderness system. A 1958 colloquy between Senator Richard Neuberger and George D. Zahn of the Washington State Cattlemen's Association exemplifies that attitude.

Senator Neuberger. Mr. Zahn, I just want to read you one sentence in the bill, because you particularly mentioned grazing because that is the interest of the Washington State Cattlemen's Association. One sentence in the bill reads as follows:

Within national forest areas included in the wilderness system, grazing of domestic livestock . . . may be permitted to continue subject to such restrictions as the Secretary of Agriculture deems desirable.

Now, do you believe that that is substantial protection for the economic interest in which you are interested?

> Mr. Zahn. Senator Neuberger, I believe that it is no protection at all. The word "may," if it was "shall," it would be excellent, but the bill in other places says existing rights if any. I believe the whole tenor for the bill is a mandate to the administering agencies of the Federal domain to place wilderness use paramount above all others.[43]

Some stockmen continued to oppose the measure almost to the end, although their protest became less strident as the bill's language was modified.

Water users formed a potentially formidable source of opposition to the bill. Many national forests owed their existence to pressure from western farmers and urbanites who wanted to protect their watersheds from damage by overgrazing or excessive timber cutting. Zahniser was undoubtedly deferring to this fact when he proclaimed watershed protection the dominant use of wilderness. But protecting a watershed and developing it so that it will yield more water in the right ways are different matters. Not surprisingly, reservoirs and waterworks were prohibited in the first wilderness bill. However, the bill's sponsors quickly realized that the water issue alone could kill the bill. For instance, Senator Thomas H. Kuchel of California, who was a cosponsor of the first bill, dropped out in the next round because of his concern over California's need for water. This problem was largely solved in 1957 when Zahniser accepted the Forest Service's suggestion that the president be allowed to authorize such projects, including the building of roads and transmission lines, when he deemed them in the national interest.[44]

Miners, like stockmen, enjoyed a long tradition of free entry on the nation's public lands. Under the Mining Act of 1872 and the Mineral Leasing Act of 1920, mining and energy concerns were given license to explore, stake claims, and apply for leases on the public domain. The miners argued that to exclude them from wilderness areas flouted American legal and cultural traditions and would damage the national interest by restricting the extraction of important metals. Although somewhat ambivalent initially, Zahniser had prohibited mining in the first wilderness bill. At the urging of the Forest Service, the bill's advocates later introduced a provision allowing mining, like water projects, when the president found it to be in the national interest. But the mining factions were not satisfied. Clearly, it would be more difficult for an individual miner or mining concern to obtain such permission than it would be for an entire community to get permission for a reservoir. Presidential authorization, as Senator Gordon Allott of Colorado put it in 1961, was "just another piece of sugar held out to make people believe that we are not going to suffer under the bill."[45]

For the next three years the bill was stalled in the House Committee on Interior and Insular Affairs as its chairman, Wayne N. Aspinall, maneuvered to incorporate congressional affirmative action and the continuation of mining. The Senate was finally forced to accept the House committee's provision that opened the national forest wilderness areas to the mining and leasing laws until January 1, 1984. This was a bitter pill for the preservationists, but according to the Senate leader of the bill, Clinton P. Anderson, the language of the mining provision precluded large-scale mechanized activity that would be incompatible with wilderness values:

> We were told that the Forest Service has managed to avoid serious damage to the primitive, wild, and wilderness areas for 25 years or more; that the Forest Service regulations governing mining activities in the areas can be continued and, indeed, that the regulations can be strengthened. The bill provides that activities in the areas shall be in harmony with the wilderness concept under reasonable regulations.[46]

Although mining interests remained opposed to the wilderness bill throughout the process, many who would have sympathized with the miners' position were placated by the compromise language in the bill's final form.

Several historians have written of the progress of the wilderness bill through the maze of hearings and past the delaying tactics employed by strategically placed opponents in the Senate and House interior committees.[47] In 1963, seven years after the bill was introduced, President John F. Kennedy and Chairman Wayne Aspinall reached a compromise providing for affirmative action in exchange for interim protection of national forest primitive areas.[48] One year later, the House adopted the measure by a vote of 374 to 1. The Wilderness Act was signed into law by President Lyndon Johnson in September of 1964.

Chapter II
The Modern Environmental Movement

"You may prefer the more facetious definition of wilderness as a place where the 'hand of man has never set foot.' Or perhaps, you'll settle for the pragmatic approach: wilderness is whatever the U.S. Congress designates as wilderness." Rupert Cutler

With the passage of the Wilderness Act in 1964, the modern environmental movement began to come of age. It had taken 8 years to get the act passed. Wilderness bills had been strongly supported in the East and by some westerners. However they had faced strong opposition from timber and mining industries and important western congressman such as Wayne Aspinall. The act was a compromise between two major interest groups that held opposite views as to how the nation's natural resources should be managed. At the extreme poles, these views took on a quasi-religious aura. Some timber industry spokesmen talked of man's duty to utilize all of God's bounty, while wilderness proponents were equally certain that wilderness was sacrosanct. The Forest Service was in the middle. As a result, the act was not a model of clarity, and it had provisions which were apparently contradictory or had been left deliberately vague. For instance, the act defined wilderness as "an area of undeveloped Federal land retaining its primeval character and influence." Two lines later this definition was qualified by the statement that on such lands "the imprint of man's work [is] substantially unnoticeable."

Was federal land that had once felt the "imprint of man" no longer eligible for wilderness designation because it logically no longer retained its primitive character, or was it eligible if that imprint was now "substantially unnoticeable"? The act did not reconcile that discrepancy.

It soon became apparent that the Wilderness Act had only set the philosophical outlines of federal wilderness policy. The rest of the picture would have to be provided by the sections of the act that require citizen involvement in the study of potential wilderness areas and affirmative action by Congress to put them into the National Wilderness Preservation System (NWPS).

During the debate on the wilderness bills, the environmentalists had strongly favored a procedure by which the executive branch would propose wildernesses that automatically would go into the Wilderness System unless Congress exercised a legislative veto. They were eventually forced to accept Wayne Aspinall's demand that only a bill which was affirmatively passed by both houses of Congress and signed by the President could create a wilderness. The environmentalists were afraid that affirmative action would make it difficult to get areas into the Wilderness System; but they soon realized that, rather than being a defeat, affirmative action was an unexpected boon because of the strength of public support for wilderness. Steward Brandborg, the executive director of The Wilderness Society, recognized this historical irony in his 1968 annual report to the society's membership.

The education and leadership training of the public, to the end that it may attend to its own interest, has been greatly aided by the Wilderness Law, particularly those provisions which were inserted by the opponents of the measure, requiring that Congress must act affirmatively on each addition to the National Wilderness Preservation System. This "blocking effort," as we saw it at the time, has turned out to be a great liberating force in the conservation movement. By closing off the channel of accomplishing completion of the Wilderness System substantially on an executive level, where heads of organizations would normally consult and advise on behalf of their members, the Wilderness Law, as it was passed, has opened the way to a far more effective conservation movement, in which people in local areas must be involved in a series of drives for preservation of the wilderness areas they know.[1]

The Forest Service was at first unprepared for the extent of public participation in wilderness designation. In late 1964 Richard Costley, the new Director of the agency's Recreation Division in the Washington Office, visited Steward Brandborg with what he later said was the naive assumption that they could quickly agree about the standards that should be used in recommending national forest areas for inclusion in the Wilderness System. Instead of reaching an agreement, Costley left the meeting extremely "shook" by Brandborg's argument that the political process should be allowed to decide the ultimate size of the wilderness system.[2] It had not always been that way. A decade earlier, a dejected leader of The Wilderness Society, Howard Zahniser, had left a meeting with the Forest Service after learning that the agency (along with the other federal land managing bureaus) would not support his first wilderness bill. The passage of the Wilderness Act in 1964 marked the end of an era in which both the Congress and the public depended entirely on the Forest Service to determine how national forest lands should be classified.

Under the Wilderness Act, Congress for the first time gave itself the power to determine how a particular piece of national forest land was to be classified. The Forest Service had a new role as one of several wilderness advisors to Congress, which now had the final word on what areas would be designated wilderness. (Subsequently Congress broadened its authority to classify national forest land with the passage of the Wild and Scenic Rivers Act and several National Recreation Area acts). As a consequence of Congress' new authority, many people in the environmental movement came to look upon the Forest Service either as a temporary obstacle on the path to enlarging the Wilderness System in a way they desired or as basically irrelevant to the process. As Ernest Dickerman, former staff member of The Wilderness Society and chief lobbyist for wilderness in the East, has said: "Everything was OK once we realized that the Forest Service can't vote."[3]

Forest Service people, on the other hand, often felt that environmentalists slighted their agency's pioneering role in administratively designating the first wilderness areas in the 1920's and 1930's and that their professional advice and leadership were not given the attention deserved. A good example of this altered relationship took place in 1973 when Chief John McGuire testified before the Senate Interior Committee on a bill to create wildernesses in the East. Senator Frank Church tactfully chided McGuire for proposing an alternative approach to eastern wilderness. Church then told McGuire that the Forest Service should stop trying to define wilderness and limit itself to making wilderness recommendations to Congress so that Congress could make the final decisions.[4]

Senator Church was being somewhat simplistic when he asked the Forest Service to send recommendations and not definitions, because it was impossible for the agency to make recommendations without some kind of implicit definition of what it was studying. Also it became clear by the late 1960's that while Congress often would enlarge agency recommendations, it would virtually never decrease them. Environmentalists might argue that the agency's recommendations were usually so small that Congress was not given the opportunity to question whether the Service had proposed too much. Forest Service personnel felt that their proposals were the "correct" size, although they assumed that whatever they proposed would be exceeded by the environmentalists and that Congress would probably designate an acreage somewhere between those amounts. In addition, the Forest Service was still charged with meeting its other congressional mandates to manage the national forests for natural resources, recreation, and wildlife.

The Wilderness Act gave some general definitions of wilderness, but it gave no guidelines on how wilderness values were to be reconciled with commodity values. That problem did not exist for the Park Service and the Fish and Wildlife Service in the Department of the Interior, two other federal agencies with large amounts of undeveloped land, because, for the most part, their potential wilderness areas already had been legislatively withdrawn from most commodity uses. The national forests, however, had been created in the late 19th and early 20th centuries when Americans were worried that the country's rapid industrial growth would be stopped by an impending shortage of raw materials. In fact, the original national forest "Organic Act" of 1897, which was not supplemented until the Multiple Use Sustained Yield Act of 1960, stipulated that the forests would be managed to provide a continuous supply of water and timber to the Nation. It did not mention recreation, scenery, or wilderness, all of which the Forest Service had to make room for under its general administrative authority until 1960. Thus, the agency had to perform a difficult balancing act in attempting to make wilderness designation compatible with its other legislated responsibilities towards the national forests. Also, if it recommended "too much" wilderness, industry would criticize it, and if it proposed "too little," it would incur the wrath of the environmentalists. It was difficult to make professional decisions in this emotional atmosphere.

What about the commodity groups themselves? They had wielded much influence during the debate over the original wilderness bills and were able to stall legislation for nearly 8 years. But that had been a national debate, and the issues had mobilized the full strengths of the contending interest groups. When it came to establishing individual wilderness areas, commodity groups labored under several handicaps. They rarely mustered the amount of public support that wilderness advocates had, and what strength they did have was not easily focused on individual wilderness areas. As John Hall of the National Forest Products Association has pointed out, every undeveloped area has some claim to ecological uniqueness, however small, that can be used to mobilize support for it.[5] Wilderness lovers in New York can be aroused to defend a threatened wilderness in Montana even if they have never seen it. Commodity interests cannot invoke the same level of far-flung solidarity. The possibility that a lumber mill might close, that some jobs might be lost, or that a fraction of a percentage of the nation's timber or mineral supply will be "locked up" does not produce the same intense public response as the image of a violated wilderness. Although a few people in industry passionately advocated their point of view, the cause of development was not as emotionally charged as that of wilderness preservation.

What strength commodity interests do have on the local level is based on fears that wilderness designation will adversely affect the number of jobs and level of income. (Environmentalists argue that in most cases this concern is exaggerated and that a few years after a wilderness is created the local inhabitants become its most vigorous supporters.) Politicians respond to this concern and often oppose wilderness recommendations that might harm their constituents' economic interests. As Brock Evans, formerly of the Sierra Club has pointed out, Congress operates under the principle of 'comity' when dealing with wilderness, i.e. bills are rarely reported out of committee unless they have the backing of the congressional representatives in whose districts or States the wildernesses lie.[6]

Consequently, a long period of local political action and education combined with outside pressure is often needed to change the political equation. This is difficult work but it was the most favored environmentalist strategy during the late 1960's and most of the 1970's. In fact, Brock Evans, one of the Sierra Club's most successful practitioners of the piecemeal approach, at first doubted the wisdom of the omnibus "Endangered American Wilderness" bill (first introduced in 1976 and passed in 1978), because he feared that it might dissipate the political strength that environmentalists could otherwise concentrate on individual areas.[7] Commodity interests, on the other hand, felt frustrated over being "nibbled to death" and over a seeming inability to prevent the slow but inevitable growth of the Wilderness System. Into this mix came what the environmentalists dubbed the Forest Service's "purity doctrine" for classifying and managing wilderness.

Immediately after the passage of the Wilderness Act in 1964, the Forest Service assembled a special task force of experienced wilderness managers to write wilderness policy and regulations in accord with the act. The group included Forest Service personnel Gordon Hammon, George Williams, Arne Snyder, Ed Slusher, and Bill Worf, as well as Bill Brizee of the Department of Agriculture's Office of The General Counsel. At first, the task force believed that they all knew what wilderness management was and therefore expected that their stay in Washington, DC would be brief. But, as they began to discuss the subject and query field personnel, they discovered that there was not a consensus nor a uniform policy. Wilderness policy was a series of regional interpretations of some general and rather flexible regulations. One of the task force members, Bill Worf, later offered his opinion that this lack of consistency demonstrated to him in a direct and personal way one of the reasons why the Wilderness Act had been passed.[8]

The task force, which shared its space in the Department of Agriculture Auditor's Building with a flock of pigeons, spent 9 weeks writing the first draft of the Forest Service Manual chapter for wilderness. According to Richard Costley, who was in charge of the project, they were given a free hand and were not subject to any external pressures. They concluded that management guidelines had to be strict if they were to be both consistent with the act and enforceable.

And the basic ingredient of successful leadership in Wilderness Management as in other unfamiliar and frequently debatable areas is consistency. This is especially significant when most of the values and criteria involved are so highly subjective in nature; when the scales of judgement are so easily tipped by emotional considerations. It soon became clear to us that if we gave different - or even seemingly different - advice in like or nearly like situations, there would be rampant confusion. We had to take clear and firm positions, demonstrate them with concrete examples, and then stick to them.[9]

The task force members were familiar with some of the most spectacular and pristine areas of the country. They believed that Congress had been thinking of this these areas when it wrote the act, and that it had charged the Forest Service with keeping them as untainted as possible. Moreover, as Richard Costley has pointed out, a Forest Service inspector needed clear and strict rules with which to evaluate wilderness managers - rules that minimize as much as possible the intrusion of modern technology and conveniences into wilderness areas.
In 1972 Bill Worf explained why he and his Colleagues had arrived at the purity doctrine:

During this entire public involvement process, other forces were also acting to shape National Forest wilderness policies. Miners were asking to use helicopters and motorized equipment for prospecting; telephone companies wanted to install electronic repeaters; NASA wanted to use helicopters and make a motorized installation for purposes that were so highly secret they would not describe them; . . . cattlemen made strong cases that bulldozers were necessary to maintain badly needed stock water ponds; Fish and Game Departments asked to use helicopters for planting game animals and fish; . . . Hardly a week went by without some new and different challenge. As each decision was considered, it was tested against the Act, the maturing policy, and other preceding decisions. These actions served to sharpen and clarify policy positions, and it became abundantly clear that specific and consistent policy guidelines were necessary to prevent gradual erosion of the wilderness resource . . .[10]

The task force also concluded that the purity doctrine also should be applied to the selection of potential wildernesses. If areas that showed the "imprint of man" were allowed into the system, the Forest Service might be pressured to allow nonconforming activities on previously undisturbed wildernesses.[11]

The Forest Service strongly believed that Congress had intended to create a high-quality Wilderness System and had written the Wilderness Act to reflect that desire. The agency knew that managing wildernesses so that they would keep their high quality would be expensive and would place some limits on the number of wildernesses it was willing to create. In December 1964, Art Greeley, then deputy chief for the national forest system, explained this reasoning to a group of regional foresters who had gathered in Ogden, Utah, to discuss a draft of the new wilderness management policy and regulations.

The Act is basically a true Wilderness Bill. It provides for no new non-conforming uses. It accepts (or tolerates) some non-conforming uses - because they are established - but provides that they shall be carried on insofar as possible so as to not destroy wilderness values.

The Act accepts that "wilderness" is expensive - an expensive luxury - a luxury which the country can afford. It is expensive. The commodity uses of the land will be foregone. The relative extra cost of conforming administrative practices, compared to some normal administrative practices that might not conform (sic). It follows that when a choice must be made between jeopardizing or enhancing a wilderness value it shall be enhanced ... But wilderness is so expensive that we probably can't afford too much more of it - either "pure" - or semi-pure.

We have 9.1 million acres - and 5 plus million headed for Wilderness. Eventually possibly 16-18 million acres altogether.

It seems we have the choice - Maybe 16-18 million acres of pure wilderness - or 2 or 3 times as much half-baked wilderness, all with an encumbrance on truly multiple-use management.

So - we should manage what we have as WILDERNESS and by so doing we won't put a semi-wilderness lien on the multiple use management of 2 or 3 times as much "possible" wilderness.[12]

Greeley's remarks show that in 1964 the Forest Service leadership considered wilderness management as separate and distinct from multiple use management, although they knew that the Multiple Use Sustained Yield Act of 1960 defined wilderness as one of the "multiple uses". They also demonstrated that the Forest Service was looking farther ahead to the possible implications of the Wilderness Act than were the environmentalists, who, until the early 1970's, were projecting a final national forest portion of the Wilderness System of about twenty million acres.[13]

The architects of the Forest Service purity doctrine, such as Dick Costley and Bill Worf, came to the principled conclusion that lax wilderness standards would result in the "cheapening" and perhaps the ultimate destruction of the Wilderness System. They also have vigorously denied that they were attempting to protect the interests of commodity users by writing strict standards. According to Costley, most wilderness advocates did not really understand the Wilderness Act or care about "any other facet of what we consider Wilderness Management ... Rather, they look upon classification under the Wilderness Act as a sure way to keep logging and/or roads out of their favorite areas."[14] And then in a statement that showed that not even veteran Forest Service professionals were immune from emotional outbursts over wilderness: "I have become 'pure'. I know it. And I know why . . . we have the difficult responsibility of the stewardship of that System. We must make sure . . . that its integrity is not violated."[15]

Costley's eloquent valedictory on his 7 years as Director of Recreation explained to a skeptical colleague why only the professional expertise of the Forest Service stood between the National Forests and a deluge of bad wilderness proposals:

We are being confronted [this in 1971] on more and more National Forest fronts with wildcat wilderness proposals . . .

The timber industry does not have the credibility or the muscle to stop these ill-advised proposals . . .

The only force which can keep wilderness classification in balance and assure future generations will continue to have wilderness is the Forest Service. And we are not going to be able to do it on the basis of the "need" for the forest products, or the roads to harvest them.

In my judgment the best opportunity - by far - for us to keep wilderness classification action sound and in balance is for us to make sure that the public comes to realize - as Congress did when it passed the Act - that while wilderness is an important part of our National heritage, it is EXPENSIVE. It is expensive not only in terms of resource opportunities foregone; it is expensive in management costs. We had to start this course by taking the basic stand that economic efficiency, or convenience, by themselves are not justifications for approving something in wilderness which is otherwise questionable. That did not go down at first, but soon the sheer overpowering logic of it began to sink in...[16]

Anticipating that there would be strong public pressure for wilderness designation, the Forest Service hoped to convince Congress to limit the amount of wilderness acreage by showing it the "true" costs involved in a pure or "strict constructionist" approach, as Bill Worf prefers to call Forest Service wilderness policy.[17] Worf and Costley foresaw that the Forest Service would encounter many, perhaps intractable, problems in protecting wilderness if high standards were not used in creating them. In other words, they did not want to be left holding the wilderness management bag for Congress and the environmentalists.

At one time many in the environmental movement also subscribed to a version of the purity doctrine - a legacy of the pre-1964 struggle to create uniform standards for a national wilderness system. In the 1930's the Forest Service had administratively set aside 14 million acres of "primitive areas." From the late 1930's until 1964 the agency had been re-evaluating them and gradually converting them administratively to either wildernesses (100,000 or more acres) or "wild" areas (5,000 to 100,000 acres). The environmentalists and the Forest Service had been engaged in a defensive struggle to keep certain activities and structures out of the primitive areas so that they would still qualify for wilderness or wild designation. Consequently, Steward Brandborg commented favorably on the first draft of the Forest Service's wilderness regulations.[18] Michael McCloskey's 1965 article on the meaning of the Wilderness Act contained language that could have been written by Costley or Worf:

Second, wilderness has been valued more as a setting for a human experience than as a source to be physically used. Only scientific research is oriented toward the resource as the subject matter. Man and the reactions his mind has to a special setting is the subject matter in all other cases. Thus, it is important to understand that wilderness is valued more as a mental image than as a physical reality . . .

Third, a strong bias against commercialism may be inferred from the nature of the interests that place value on wilderness. These interests view wilderness areas as geographical secessions from the American economy. Recent trends are strongly toward making the secessions complete.[19]

The environmentalists' position began to change when they realized that the Forest Service wanted to use the same strict standards to recommend wildernesses as it did to manage them. Their organizations were no longer simply trying to preserve the status quo but were now attempting to enlarge the Wilderness System. New roles created new perspectives for both the Forest Service and the environmentalists. The Forest Service moved consciously to a more pure position when it went on the defensive. The environmentalists underwent a reverse evolution when they took the offensive.

The environmentalists found nothing in the Wilderness Act that required identical standards for management and allocation, nor would they accept the argument that the need to apply certain management techniques justified pure standards. They charged the purity doctrine was applied selectively when the Forest Service wanted to exclude an area from the Wilderness System, usually for economic reasons. Industry representatives occasionally came out with tongue-in-cheek caricatures of rigid purity but their views were given little weight on the subject because, as Costley noted, "many of those who fall into this category do not support the philosophy that the maintenance of an enduring wilderness resource is in the public interest, and many would like to undermine the wilderness concept."[20]

The National Park Service, in an attempt to preserve some administrative freedom, also fell back on an implicit purity argument in order to exclude from its wilderness recommendations small areas containing structures and improvements in otherwise roadless areas, thus creating a "Swiss Cheese" effect. In 1972 Douglas Scott, then on The Wilderness Society staff in Washington, drafted a speech on this subject for a congressional hearing. The hearing was chaired by Senator Frank Church, the floor leader of the 1964 Wilderness Act. Using Scott's text, Church berated an assistant secretary of the Interior, Nat Reed, who was a willing accomplice to his own public chastisement.[21] These proceedings essentially put an end to the Park Service purity doctrine.

The Forest Service also came under increasing pressure in the early 1970's from congressional leaders, such as Senator Church, to jettison its purity doctrine. It felt more strongly about purity than the Park Service because it had many people with a strong philosophical commitment to the doctrine and because its leaders felt that it was necessary in order to maintain a "proper" balance between wilderness and other multiple use values.

As mentioned earlier, the Wilderness Act defined wilderness somewhat ambiguously, and its legislative history was so long and complex that it is impossible to extract an unambiguous meaning from it. Bill Worf, who enjoys the respect of many environmentalists as a worthy and principled sometime opponent, maintained that the act implied that wilderness was a "non-renewable" resource. A Forest Service colleague countered that:

> *Nature is a great healer and in just a few years imprints of man's work will become substantially unnoticeable. As ecologists, we can even forecast approximately when the imprints will become insignificant. The argument is made that Wilderness, by definition, can never be restored. In actual fact, there are many examples where, for all practical purposes, Wilderness has been restored. It is just a matter of how much time is involved.*[22]

The timber industry picked up this argument during the debate on the eastern wilderness bills of the early 1970's, and suggested that if areas could be restored to wilderness condition then they could be rotated along with timber harvests. However, this reasoning was not acceptable to wilderness enthusiasts who had developed attachments to specific places and were unwilling to accept substitutes during the decades it would take to restore them. Wildernesses may be renewable but they are not substitutable, because, as McCloskey pointed out, wilderness lovers often value the "mental image" of a wilderness more than its actual physical reality.

Ernie Dickerman maintains that because the three principal drafters of the first wilderness bill (Howard Zahniser, Harvey Broome, and George Marshall) were easterners who had their first wilderness experiences in eastern forests, they could not have written a definition that would have excluded these areas. On the other hand, he has also admitted that it took nearly 8 years after the passage of the Wilderness Act for environmentalists to push for an eastern bill, because at first many of them were not sure that the cutover areas in the East qualified as true wilderness.[23]

Ecologists have pointed out that it is difficult to define a "pure" wilderness because virtually all of the forested areas of the United States have experienced some form of human impact. Although most western wildernesses have been free of timber harvesting and are substantially without roads, for nearly 60 years the Forest Service was successful in reducing the number of forest fires in them. The long-term exclusion of fire may have caused as much ecological change in the drier, slow-to-recover areas of the West as timber harvesting and roads in the humid, fast-growing forests of the East.[24]

Not having any definitive statutory principles and very few examples of completely unmodified wilderness to go by, the Forest Service has realized that Congress must continue to define what qualifies as wilderness through the bills it passes and the legislative reports it writes. In the following pages we will examine some of the important events which have led us to the present situation.

CHAPTER III
Primitive Area Reviews and The Parker Decision

In order to pass the Wilderness Act, wilderness advocates had reluctantly agreed to accept affirmative action by Congress in exchange for interim protection of the Forest Service's primitive areas. The act required the Forest Service to study these areas and report its recommendations to Congress within a 10-year period. In 1963 Wayne Aspinall had inserted a clause in the wilderness bill that prevented the President from enlarging a primitive area by more than 5,000 acres. John Saylor countered with an amendment permitting the President to recommend to Congress that undeveloped land contiguous to primitive areas also become part of the Wilderness System.[1] Aspinall did not attempt to block Saylor's amendment because it upheld the principle of affirmative action by Congress. Although Saylor introduced this amendment because he knew environmentalists were interested in some contiguous areas, neither he nor Aspinall could have suspected the full consequences of its seemingly innocuous language - a national forest portion of the Wilderness System much larger than 14 million acres and a major legal defeat for the Forest Service as a result of the Parker Decision of 1970.

The Forest Service completed all of its reviews within the 10-year period. Most of them were conducted without major conflict with commodity groups or environmentalists but a few were marked by controversy as both environmentalists and the Forest Service attempted to assert their interpretations of the definition of wilderness as stated in the Wilderness Act of 1964.

The San Rafael Primitive Area, established in 1932, consisted of nearly 75,000 acres on the Los Padres National Forest on the central California coast. The area was noted for its grassy balds known as *potreros*, scenic rock outcrops, mountain lions, bears, eagles, condors, and Chumash Indian rock paintings. In 1935 Forest Service Chief F.A. Silcox assured the National Audubon Society that a proposed fire road in the area would be closed to the public so that condor nests would not be disturbed.[2]

Forest Service leaders chose to make the San Rafael its first study area because they thought it would be relatively "easy."[3] There were no timber or mineral resources to speak of in the area and while farmers and residents of Santa Barbara depended on streams originating in the area's watershed, the Forest Service correctly felt that there would be little problem working with these groups.

The first hint that things would not go as smoothly as the Forest Service had hoped came in a December 3, 1965, letter from Michael McCloskey, then conservation director for the Sierra Club in San Francisco, to Secretary of Agriculture Orville Freeman. McCloskey conveyed the Santa Barbara Chapter's complaint that the Forest Service was lobbying the Santa Barbara City and Water Commissions to support its proposal for a 110,000-acre San Rafael Wilderness.[4] Deputy Forest Service Chief Art Greeley replied to McCloskey that the Forest Service had to defend its position in public forums and that the only question was "how ardent an advocate should the Forest Service be for something that is a Forest Service proposal."[5] Regional Forester Charles Connaughton told McCloskey that he was "astonished" that the relations between the Sierra Club and the Forest Service had sunk to such a low level.[6] He requested a meeting with the president of the club. Apprarently that meeting had a salutary effect because for the next year compromise and cooperation were the watchwords for both the Forest Service and the environmentalists. The Forest Service increased its proposal to 143,000 acres and was praised by Steward Brandborg for its constructive attitude.[7]

In 1967 the temporary peace began to unravel. The sticking point was 4,200 acres of *potreros* on the Sierra Madre Ridge that the Forest Service wanted to exclude because it felt that a road and an administrative site there were incompatible with wilderness designation and because it wanted to use the area as a fire break. The Forest Service had already converted some of the brush to grass and planned to do more "type-conversion" work as soon as the San Rafael Wilderness bill had been passed and signed. The agency maintained that this disputed area was the only suitable site for a fire break, which was essential to stop the kind of fast-moving, devastating fires which were endemic to the region.

The local environmentalists maintained that the land was of wilderness character and should be protected because it was a condor flyway and contained Chumash Indian pictographs. They disputed the need for a fuel break and cited the authority of the Los Padres' fire boss who questioned the efficacy of a fire break in that particular location. The Forest Service replied that it had closed the Sierra Madre Ridge Road to the public and that it could easily protect the condors and pictographs without having to put the area into the Wilderness System. According to Bill Worf:

> When we learned of the wilderness folks' concerns, I was assigned to work directly with the Forest Supervisor to see if we could resolve the differences. The Forest Supervisor provided us in the Washington Office with large scale aerial photos of the disputed area. After much discussion the Supervisor agreed to forgo any plans to expand the already existing fuel breaks on the wilderness side. Accordingly we used the photographs, and drew a revised boundary on the edge of the fuel breaks that had already been completed. These fuel breaks had been created by bulldozing and burning the chaparral and then plowing and seeding to grass. We felt that even under very liberal interpretation these areas could not be considered wilderness. The dozer and plow marks were clearly visible on the aerial photos.[8]

Rupert Cutler, then assistant executive director of The Wilderness Society in Washington, DC, went to Santa Barbara with the hope that he could break the impasse. He and the Forest Supervisor viewed the San Rafael by helicopter because the roads were closed by snow; this action was frowned on by some local environmentalists. When he returned to Washington, DC ready to reach a compromise, he was surprised to discover that the Santa Barbara Sierra Club already had convinced Senator Thomas Kuchel to introduce a bill calling for 158,000 acres, including the 2,200 acres on the Sierra Madre Ridge that the Forest Service wanted as a fuel break. Cutler's attempt at mediation had been aborted, and The Wilderness Society's leadership now felt compelled to support the local citizen initiative.[9] Here in the first primitive area bill was a strong signal that affirmative action meant that questions of wilderness allocation would not be decided exclusively at an executive level in Washington, DC. It was also an indication that a rough division of labor was developing within the wilderness movement. The Sierra Club claimed California and the Pacific Northwest as its territory, while leaving the Rocky Mountain states pretty much to the The Wilderness Society.[10]

Senate hearings were held on the Kuchel bill. The environmentalists agreed to drop 13,000 acres from their proposal but would not budge from their insistence that the 2,200 acres on the ridge be included in the wilderness. The Senate accepted the Forest Service's recommendation and passed a bill that excluded the disputed area. It was referred to the House Interior Committee where John Saylor was able to restore the acreage. An angry Chairman Aspinall commented on these developments and implicitly acknowledged the unforeseen consequences of affirmative action and the Saylor amendment on contiguous areas.

> This primitive area was 74,990 acres. I repeat: 74,990 acres. After the first hearing, this was raised to 110,403 acres. Then, after the final hearing, it was brought to the Congress with 142,918 acres . . . The gentlemen from Pennsylvania succeeded in getting 2,200 more acres, which makes a total of 145,118 acres, . .

> May I say to my colleagues, if this is going to be the trend in our determination of whether or not primitive areas are to become wilderness areas, and if we are to increase them by 100 percent, then my opinion is that creation of new wilderness areas in the future are going to be very few and far between, because what we proposed in the first place was an additional 5 million plus acres of land, which were then primitive areas to be incorporated, . . .[11]

When the House passed the Saylor-amended bill on October 1, 1967, the lines had been rigidly drawn. The environmentalists were fearful (it later turned out unnecessarily) that if Congress acceded to Forest Service wishes in the San Rafael case, it would forever after "rubber-stamp" Forest Service wilderness proposals. The Forest Service, on the other hand, felt that the very basis of its professional reputation was at stake. If an agency that was identified in the public mind with Smokey Bear could not be trusted to make authoritative recommendations on fire management, what could it be trusted to do? A mere 2,200 acres had assumed a symbolic political importance far beyond their intrinsic worth as wilderness.

According to Congressman Baring:

> *I would like to emphasize that this is not a question of the inclusion or exclusion of 2,200 acres of land. It is whether or not the Forest Service, the agency responsible for fighting fires, will be permitted to establish a fireline at the location its experts maintain is the logical and best position to control a fire. I strongly feel that in matters such as this, the position of the agency concerned with fire prevention must be given weight. In the final analysis they are the ones that must commit men and equipment to the fireline and they are the ones responsible for the safety of the firefighting crews as well as the protection of public and private property from destruction by fire. . . . The issue is one of emotionalism versus professionalism.[12]*

John Saylor replied that:

> *The Forest Service simply does not want to see its proposal changed by the Congress in response to the conservationists' testimony. The Washington headquarters staff of the Forest Service, trying to run this Nation's public forests as though they were European forestmasters instead of public servants, have dictated their San Rafael boundaries to us, and we are expected to accept them without question.[13]*

A House-Senate conference committee decided to drop the 2,200 acres from the bill. Morris Udall had been on the conference committee but had delivered his proxy to another member who accepted the Senate's version. Saylor and Udall attempted to have the bill committed to conference but were defeated by voice vote on the House floor. On March 21, 1968, President Johnson signed the bill placing the first primitive area into the Wilderness System. Inadvertently rubbing salt into the Forest Service's wounds, President Johnson turned to Interior Secretary Stewart Udall, who had not been involved with the bill, and said, "Good work, Stewart."[14]

Both the environmentalists and the Forest Service had lobbied extremely hard for their proposals. The Forest Service's campaign had been so vigorous that after it was all over Wayne Aspinall lectured Reynolds Florance, the agency's director of Legislative Affairs, that he never wanted to see such strong advocacy again. To drive home his point, Aspinall punctuated his words with periodic finger jabs to Florance's midsection.[15] John Saylor was incensed and vowed that for every acre of wilderness lost in the San Rafael he would get 100,000 acres somewhere else.[16] He immediately got to work on a bill to create a North Cascades National Park in Washington State by transferring some Forest Service primitive land to the Park Service. To Associate Chief John McGuire, the environmentalists' tenacity on 2,200 acres was the first indication "that they were out to get as much acreage as possible."[17] The Forest Service won on San Rafael but it was probably a pyrrhic victory for it motivated the environmentalists and their congressional allies to work harder on later wilderness initiatives.

One of the next primitive area proposals, the Mount Jefferson, showed that the San Rafael had not established a precedent and that Congress was willing to take a stance independent of both the Forest Service and the environmentalists. The Mount Jefferson Wilderness, located in the Williamette National Forest of Oregon, was to be a thin segment of the scenic Cascade Crest. The Forest Service proposed a 97,000-acre wilderness. The Wilderness Society pushed for 120,000 acres, including the 3,000-acre Marion Lake exclusion and some land that had been partially logged over. The society contended that this land had been deliberately logged a few years before to keep it out of the Wilderness System. The Forest Service denied any such sinister motive, contending that it had been logged following established plans. Marion Lake was a semideveloped recreation site for boaters and fishermen containing an administrative site, campground, and boat storage facilities. It penetrated 3 miles into the proposed wilderness. If it were excluded there would be a gash in the Mount Jefferson and the possibility of further recreation development which would adversely affect the sense of isolation in the surrounding wilderness. The Forest Service argued that the heavy use and development at the lake precluded its consideration as wilderness. According to historian Ronald Strickland:

> *It was evident after the hearings that a strong case had been made both by the Forest Service and by those who advocated Wilderness status for Marion Lake. A discrepancy had appeared between the existing pattern of recreation use and Wilderness Act standards of what a wilderness experience should be. If the lake were included in the wilderness, it would not be, according to Chief Cliff, "an untrammelled area. It is being heavily trammeled." [Untrammeled was one of the adjectives used in the Wilderness Act to define wilderness.] If it were not included, the new Mount Jefferson Wilderness would be unnecessarily narrow and incomplete.[18]*

Congress excluded the logged-over area but not Marion Lake, and its committee report directed the Forest Service to remove the recreation facilities and boats and restore the wilderness character of the area. This was interpreted by the Forest Service as a sign that Congress agreed with its strict definition of wilderness.[19] (Several years later the Senate Interior Committee felt it had made a mistake in giving the Forest Service such explicit management direction.)[20] During the next several years the Forest Service followed congressional direction and removed the boats and facilities. It granted several delays to boat owners but in the process stirred up some consternation among the recreationists who had used the area.

Environmentalists sometimes pointed to Mount Jefferson as an example of the agency's "vengeful" application of the purity doctrine but this charge was unfounded because it had scrupulously followed the congressional mandate.[21] The fact that these wishes were expressed in a commitee report rather than directly in the legislation made it appear to some that the Forest Service was acting on its own.

One of the most interesting primitive area controversies took place over the DuNoir Basin next to the Washakie Wilderness in northwestern Wyoming. Here for the first time jobs, profits, and the purity doctrine were openly intermingled. The DuNoir Basin in the Wind River Ranger District of the Shoshone National Forest was directly west of the Stratified Primitive Area. In 1966 the Forest Service proposed to join the Primitive Area with the South Absaroka Wilderness and call the combined area the Washakie Wilderness. The agency left out the DuNoir Basin drained by the East and West DuNoir Creeks and some other smaller areas below the volcanic escarpments of the Stratified Primitive Area.

During the 1960's there had been extensive timber harvesting in the Wind River Ranger District. The biggest customer for national forest timber in the region was the U.S. Plywood lumber plant in DuBois, which employed over 120 workers and was operating at only about 50 percent of installed capacity. The DuNoir Basin was the only valley in the Upper Wind River Region that had not been harvested for timber. It contained approximately 100 million board feet of timber, of which at least 30 million was merchantable.[22]

The Forest Service excluded the DuNoir because it wanted to offer its timber for sale and to develop recreation sites in the Basin. The Teton and Yellowstone National Parks had limited the number of campgrounds they intended to build to maintain a planned experience level in the parks. The tourist overflow from these extremely popular National Parks was pressing on the recreational capacities of the neighboring National Forests. Complicating this situation was the fact that between 1921 and 1929 parts of the Basin had been selectively cut for railroad ties.[23] According to a 1981 Forest Service report, 30 to 40 percent of the trees on 11,000 acres of the 28,800-acre area had been cut during that period.[24] Stumps, a few skid roads, which had become jeep trails, and the rotting remains of the tiehack camps showed the "imprint of man's" work.

Did this disqualify the area as wilderness? The Forest Service, U.S. Plywood, the timber industry, the Governor of Wyoming, and the State stockmen's associations thought it did. Wilderness organizations, outfitters, many of the citizens of DuBois, the Upper Wind River Cattlemen's Association, and the Wyoming Game and Fish Department thought it did not. The Game and Fish Department believed that roads providing greater public access to the area would disrupt the herd and force it to go elsewhere. The Forest Service, on the other hand, contended experience had shown that elk could live with development and would willingly cross roads.[25]

The principal advocates for the enviromentalists were Clifton Meritt of The Wilderness Society, Orrin and Lorraine Bonney of the Sierra Club, Tom Bell, editor of the High Country News in Riverton and a Wilderness Society cooperator, and William Crump of the Game and Fish Department. Arrayed against them were Governor Hathaway, the manager of the U.S. Plywood plant, and in varying degrees, officers of the Shoshone National Forest. Senators Gale McGee and Clifford Hansen and Wyoming's Representative-at-large, Teno Roncalio, had to make the political decisions in this contentious and sometimes rancorous atmosphere.

In August 1966 a Sierra Club team led by Orrin and Lorraine Bonney of Jackson Hole, Wyoming, inspected the DuNoir Basin and the Stratified Primitive Area. They reported on the unsightly condition of most of the clear-cut land in the Wind River Ranger District and urged that the DuNoir be given wilderness protection as an elk habitat and as an accessible undeveloped gateway to the proposed Washakie Wilderness.

There is a minimum evidence of a few select trees being taken for saw timber along the ridge above West DuNoir Creek. A jeep road runs south from Murray Lake . . . This jeep road could be closed and the forest would soon revert. The upper DuNoir is better forested and the lumbering threat is greater than any other section we saw in the Stratified Primitive Area . . . Fast action will be necessary to save this part of the country as wilderness.

. . . There are about 2,500 acres of potentially valuable timberland in the extension, mostly in East DuNoir drainage. These areas were logged off at the end of the last century, and the effects are still evident. The quality of the new growth is uneven, though the forest here has made a significant beginning toward recovery of its original character... East DuNoir drainage serves as a migration route for elk from the South Absaroka Wilderness Area to the East Fork Elk Winter Pasture on Bear Creek, one of the major reasons for limiting the region to wilderness usage . . .

As the largest timbered area in the contemplated Stratified Primitive Area, the two drainages would be vital to reproducing in microoocosm the ecology which once characterized the Rocky Mountains, or to be more exact, the latter day derivative of that ecology.[26]

The environmentalists privately acknowledged that the Forest Service was in a very difficult position on the DuNoir. In 1969 Tom Bell wrote to one of his collegues that District Ranger:

[Harold] Wadley is under tremendous pressure from U.S. Plywood. The previous ranger had yielded to pressure and the district has been badly overcut. Wadley cut timber sales from about 17 million board feet last year to 3.8 million this year. He told Senator McGee and me that he doubted if the District could now sustain timber yields of 2 million board feet per year from here on out. Nevertheless, Plywood is threatening him - and could eventually get him transferred. He is not a patsy for anyone. We are not going to stampede him nor is U.S. Plywood. But he is trying to do the right thing and I think we should support him.[27]

Shoshone National Forest Supervisor, Jack Lavin, defended the Forest Service's proposal at the public meetings but when questioned about other areas that had gone into the Wilderness System despite evidences of past human activity, he replied that the final decision rested with Congress and that the Service would abide by that decision.[28]

From his position as wilderness staffer in the Washington office, Bill Worf enumerated to William Isaacs of the Wyoming Wildlife Federation six reasons why the DuNoir should be left out: (1) the Wilderness Act provided only one set of criteria for both allocation and management and that if areas like the DuNoir went into the system, the Forest Service would be pressured to ease up on its management restrictions; (2) the 679,520 acres in the Forest Service proposal was "an adequate Wilderness unit"; (3) the Forest Service had to recognize other resource values such as timber and family recreation; (4) over 52 percent of the Shoshone National Forest was already in the Wilderness System; (5) both the Park Service and the Forest Service had pending wilderness proposals in other parts of western Wyoming; and (6) "some hunters and some outfitters prefer that elk hunting country remain unroaded. I, too, enjoy my hunting much more where the folks in jeeps cannot get, but it doesn't follow that Wilderness status is necessary to maintain an elk herd."[29]

The DuNoir convinced many environmentalists that the purity doctrine was not so much a consistent philosophical proposition as a practical attempt to limit the amount of wilderness acreage. According to Tom Bell:

It seems to me that such a definition [in the Wilderness Act] leaves much latitude. Congress must have intended it that way for some of the very first areas included within the system had selected cut-over areas within the wilderness (Bob Marshall Wilderness area, Montana, being the best example). Only the past week Congress included the Desolation Wilderness in California which has within it two dams and reservoirs. After the Bridger Wilderness was included in the system, a steel bridge was brought in and placed by the Forest Service for the convenience of people. There is a mining ghost town high in the Big Horn Mountains and well within the Cloud Peak Primitive Area [in Wyoming]. We are trying to ask for its inclusion and I am certain the Forest Service will also.

So what is substantially unnoticeable? Or untrammeled by Man? How selective do you get?

You can easily see that on certain areas where the Forest Service doesn't want to include an area, or where it is under pressure not to do so, it is very easy to fall back on the purity of the area in question . . .[30]

By 1969 Senator Hansen had decided that the DuNoir should be excluded from the Washakie Wilderness, although he opposed immediate timber harvesting. Senator McGee and Congressman Roncalio supported the inclusion of the Basin. Roncalio was especially adamant that the lingering signs of human activity in the DuNoir did not disqualify it as wilderness. During an inspection trip to the area, he underscored this point by kicking an old stump and exclaiming as it disintegrated under his feet: "So much for the imprint of man."[31]

In August 1970 Hansen and McGee reached a compromise on the Washakie proposal. Hansen agreed to accept additional acreage in the Ramshorn area near the DuNoir Basin in exchange for McGee's promise to support an amendment placing the DuNoir in a special management unit in which timber harvesting and further road construction would be prohibited. In addition, the Forest Service was to study the Basin and give Congress its recommendations within 5 years. Wyoming conservationists agreed to accept the compromise after being told by McGee that it was essential to get the bill out of the House and Senate Interior Committees.[32] Clif Merritt protested to Tom Bell that Wyoming conservationsists had been close to their goal of including the DuNoir and that a special management unit could establish a bad precedent that could be used to exlude other areas from the Wilderness System.[33] (The Forest Service also was not pleased with the concept because of a concern that it might lead to more legislative "zoning" of the national forests.) Merritt's arguments convinced Tom Bell and a month later he informed Hansen that the Wyoming Outdoor Coordinating Council, of which he was executive director, was withdrawing its support from the compromise.[34] Aware that the compromise was still supported by other Wyoming conservationists, such as the Bonneys, Hansen replied to Bell that he would continue to advance it.

I feel that I could take exception to several points in your letter, but I think it will be sufficient to say that I realize that there are others besides conservationists who feel that the boundaries are not precisely right and that some areas should have been deleted, just as you believe some should have been added. Likewise, for many the restrictive language controlling the DuNoir is not acceptable.[35]

The Senate passed the compromise bill. Roncalio's strong advocacy in the House succeeded in getting the DuNoir in its bill, but the Senate-House Conference committee restored the compromise provision. The Washakie Wilderness was signed into law in October 1972. During the next several years the Forest Service studied the DuNoir. At first the agency recommended that 11,000 acres at the higher elevations be designated as wilderness and that the rest be released for other uses. When Rupert Cutler became assistant secretary of agriculture for conservation, research, and education in April 1977, he directed the Forest Service to be less restrictive in evaluating evidences of past human activity in wilderness studies. On April 1, 1978 the Forest Service held a public hearing on its proposal in Dubois. Sixty-three of the eighty participants spoke in favor of wilderness designation for the area. As a result, the Service decided that the entire DuNoir basin was worthy of wilderness protection. In 1978 Roncalio sponsored a bill adding the 28,800-acre Special Management Unit (plus about 5,000 acres of contiguous roadless land) to the Washakie Wilderness. He was critical of Louisiana Pacific's purchase of the U.S. Plywood plant in 1974 and refused to help them out of what he thought was a bad management decision by opening all or part of the DuNoir to timber harvesting. Senator Hansen refused to support the bill, arguing that Roncalio had already gotten the 15,000-acre Savage Run Wilderness into the system (as part of the Endangered American Wilderness Act) during that congressional session.[36] At the time of this writing the fate of the DuNoir is still being debated, although it is protected from development unless and until Congress determines otherwise.

The Parker Decision, discussed next, was an important episode in the wilderness movement. During the 1950's and 1960's wilderness organizations did not resort to the judiciary. They knew that the courts adhered to the principle that the "sovereign" was immune from unconsented law suits against its administrative actions and that private individuals or organizations could not gain "standing" to sue unless they could prove that they had suffered or could suffer personal economic injury.

Beginning with the landmark Scenic Hudson decision of 1965, where the judge found in favor of local ad hoc conservation organizations opposing a proposed hydroelectric project that was to be licensed by the Federal Power Commission, the courts began to re-define and liberalize the conditions under which the federal government could be sued.[37] In 1969 wilderness organizations had their first opportunity to test these recent judicial precedents.

East Meadow Creek was a largely undeveloped area directly west of the Gore Range-Eagles Nest Primitive Area on the White River National Forest in north central Colorado. It was about 9 miles north of the ski resort town of Vail which had been built in 1964. Like the DuNoir Basin, East Meadow Creek was not too high, rugged, or inaccessible for the ordinary hiker. According to Rupert Culter writing in his 1972 doctoral dissertation, East Meadow Creek

functions as the gateway to that popular backpacking, horseback-packtrip, and big game hunting area for wilderness travelers who begin their trips at the ski resort town of Vail. The east Meadow Creek drainage is rolling, timbered high country at a 9,200 to 15,000 foot elevation. Small meadows and park-like stands of old-growth Englemann spruce, lodgepole pine and fir contrast there with dense thickets of young lodgepole pine and fir, the aftermath of fire. The area's claim to importance as a fish and wildlife habitat is based in part on its role as an elk migration route and nursing ground.[38]

In 1962 the Forest Service drew up a plan to log East Meadow Creek, and 2 years later built an access road to the border of the area. A low-standard truck-trail had been constructed in the area in the 1950's in an effort to combat a bark beetle infestation. After the passage of the Wilderness Act, the regional office in Denver sent an investigator to see if the area should be added to the primitive area. He recommended that it be excluded because of the "bug road," which was used by motor vehicles, the existence of some private inholdings and unpatented mining claims, and because it fell "outside the ridge top hydrographic divide, a recognizable natural feature to serve as the wilderness boundary for ease of identification and administration."[39]

The White River Forest Supervisor and the Eagle District Ranger favored logging East Meadow Creek to provide sufficient timber "for established sawmill operators." The regional office had to approve the sale because it exceeded the 5 million board feet that the supervisor could authorize on his own authority. The regional office faced a dilemma because the primitive area review had not been completed; and, if it approved the sale, it could be accused of "trying to control wilderness classification by timber harvesting."[40] Regional Forester David Nordwall decided that only a few of the proposed timber blocks would be put up for sale and that a buffer zone would be left between the sale area and the boundary of the primitive area. His compromise did not work.

Once again, however, the Forest Service was caught in the crossfire between two potent interest groups. Both groups claimed that the Forest Service was usurping the prerogative of Congress to establish wilderness boundaries. Wilderness proponents saw implementation of any or part of the timber sale as effectively reducing the size of the area which could be recommended to Congress for Wilderness Act protection. On the other hand, the forest products industry saw the Regional Forester's decision to postpone the sale of eight of the fourteen cutting blocks in the East Meadow Creek unit as an unauthorized expansion of the primitive area's boundary prior to any action by Congress sanctioning the closure of this national forest land to logging.[41]

Citizens of Vail were especially agitated. They argued that the sale had been planned in 1962, before the establishment of their town, which depended on recreation dollars for its existence. Some of them approached Clif Meritt in Denver, who told them that an injunction was the only solution. He suggested they contact Tony Ruckel, a young environmentally minded criminal lawyer (he now works for the Sierra Club Legal Defense Fund) who had recently moved to Denver from Washington DC, where Merritt had known him.[42] Ruckel agreed to take the case for a modest fee, while freely contributing some of his own time. Court expenses, including a $10,000 bond required by the court, were paid largely by the Sierra Club, which, along with 12 citizens of Vail, and several Colorado conservation organizations, was one of the plaintiffs in the case.

The conclusion was far from predictable when Ruckel filed a motion for a preliminary injunction before District Court Judge William E. Doyle in April 1969. Merritt had tried to persuade The Wilderness Society to join the suit because if "successful, this case will set a great precedent in requiring complete re-evaluation of Forest Service practices in this region."[43] The Society declined Merritt's offer because it feared the case would drain its limited financial resources and because it had received advice such as the following from a Washington lawyer, who, like many of his colleagues, had not yet assimilated the implications of the Scenic Hudson decision.

It is the law that the United States cannot be sued without its consent . . . On this principle the Government is on sound ground in moving to dismiss.

The principles listed above are elementary. I do not think the plaintiffs have a chance. They may succeed in getting some publicity and in interferring with the inevitable course of events.

If this is their purpose, which cannot be the case, the action is an abuse of judicial process.

In this last sense, ignorance of the law may be a defense.[44]

At first Ruckel thought that he could base his case on the Multiple Use Sustained Yield Act, but Merritt convinced him that the act's language was too broad to be used as the principal argument. He (and other environmentalists) directed Ruckel to the Wilderness Act which said "nothing herein contained shall limit the President in . . . recommending the addtion of any contiguous area of national forest lands predominately of wilderness value."[45] At the preliminary hearing Judge Doyle told Ruckel that if he had a case it would be based on that provision of the Wilderness Act. According to Merritt, "from that time on that was the only issue."[46]

Judge Doyle was at first skeptical of the plaintiffs' arguments but gradually came around to accepting them.[47] He upheld their right to sue the federal government, accepted the evidence presented by "star witnesses" Clif Merritt and Bill Mounsey that East Meadow Creek was "predominantly of wilderness value" (despite Forest Service arguments that the "Bug Road" disqualified it), and interpreted the language of the Wilderness Act to mean that the Forest Service must refrain from developing a contiguous area which was potentially of wilderness value until the president and Congress had acted on the agency's recommendations. On February 17, 1970, Judge Doyle found in favor of the plaintiffs and permanently enjoined the timber sale. On October 1, 1971, the Tenth Circuit Court upheld Judge Doyle.[48]

The Forest Service and government attorneys thought "bad" law had been made because they believed the Wilderness Act did "not require review of every contiguous acre of land of wilderness character by the President and Congress but does not prohibit such consideration." Russell Train, the director of the Environmental Protection Agency, proposed that the President issue an executive order affirming the validity of the circuit court's decision. The Forest Service successfully opposed Train's suggestion.[49] The Department of Justice appealed to the Supreme Court, which refused to hear the case.

The agency realized that technically the decision only applied to the Tenth Circuit. However, it was clear to many in the Service that they would have to bear the decision in mind whenever they wanted to develop lands contiguous to primitive areas anywhere in the national forest system. Along with the Lincoln-Scapegoat situation, to be discussed in the next chapter, the Parker Case (named after one of the plaintiffs) pushed up the timetable when the Forest Service would have to confront the general issue of the future management of its millions of acres of undeveloped forest land. Confident that contiguous areas they were interested in would be protected, the environmentalists could shift their attention to other types of undeveloped national forest land.

Chapter IV
The Lincoln-Scapegoat - The First De Facto Bill

The primitive area reviews and the Parker Case involved land already protected by the Forest Service or contiguous to such land. But there were millions of other undeveloped acres of Forest Service land that were either contiguous to established wilderness areas or detached. Neither were covered by the primitive area reviews or the Parker Decision. Environmentalists called these "de facto" wilderness, a term that the Forest Service usually avoided because of the implication that de jure status was just around the corner. In fact until World War II perhaps as much as two-thirds of the national forest system was essentially undeveloped.

For forty years, from 1900 until 1940, the administration of the national forests in the west was pretty much a job of custodianship. These were the days of the pack-string, the lookout towers, and the isolated fire guard stations. The national forests were de facto wilderness, largely unaltered from their primeval conditions, seldom visited by man, and the "hard-rock" forest rangers came to hold a deep affection for this wild uninhabited country. But as World War II approached and arrived, a demand for timber products made a lasting impact on the national forests. The Custodial Era began to fade as the Management Era dawned; logging trucks and power saws thundered in the woods, timber production climbed gradually during the war and skyrocketed after it. The long pent-up civilian demand for houses (among other things) produced a building boom of heroic proportions.[1]

The first wilderness bills had been put forward primarily to prevent the development forces of the "Management Era" from encroaching on the Forest Service's primitive areas. Some environmentalists anticipated that other undeveloped parts of the national forest system would need legislative protection. The Wilderness Act of 1964 said nothing about de facto wildernesses in the national forests, although it required two Interior Department agencies to study their roadless areas for possible wilderness designation. The statutory authority for studying de facto national forest areas was contained in the Multiple Use Sustained Yield Act of 1960, a piece of legislation that the environmentalists had, for the most part, resisted. The law stated that the "establishment and maintenance of areas of wilderness are consistent with the purposes and provisions of this Act."

The Forest Service 1964 Wilderness Task Force recognized that de facto areas eventually would have to be considered but the agency's leadership hoped to wait until at least 1974, the year the primitive reviews were to be completed, before it began a formal study of these areas.

"Precedent" is often a cliche word when used outside its technical meaning in the law. In the early years after the passage of the Wilderness Act it was frequently uttered by agency personnel charged with interpreting the act and anticipating congressional intentions. It was applied to the Lincoln-Scapegoat controversy, which was not an exaggeration given the impact this controversy would have on wilderness politics.

The Lincoln Back Country was originally an area of 75,000 acres of undeveloped forest land in the northern half of the Lincoln Ranger District of the Helena National Forest, 12 miles north of the town of Lincoln in the northwestern part of Montana. To its north, separated by the Scapegoat Mountains, lay the Bob Marshall Wilderness, the "jewel" of the Forest Service's wilderness system. Scenically undistinguished from "literally millions of similar acres in western Montana,"[2] the Lincoln Back Country was nevertheless an important hunting, fishing, and hiking area for people converging on it from Missoula, Butte, Helena, and Great Falls. Called the "Poor Man's Wilderness" because it was easily accessible to day hikers and did not require the services of an outfitter, it had been a favorite camping area of Clif Merritt, who became The Wilderness Society's western regional representative in 1965. The Lincoln Back Country, which had been named by its longtime booster, Cecil Garland, was not formally protected by the Forest Service. However, a sign at an entrance to the area prohibited the use of motorized vehicles, which, according to Merritt, led many people to believe that it was a wilderness.[3]

Forestry professor R. W. Behan of the University of Montana stated in 1965 that the Back Country's timber resources were meager: "isolated patches of larch-fir type and a scattering of sawtimber of spruce are found in the vicinity of Heart Lake and in the Meadow Creek watershed respectively, but the negligible values involved could not, under foreseeable cost-price conditions, justify the investment in access-roads necessary to reach them."[4] But what it lacked in timber and scenery was compensated for by an abundance of game and fish. For instance, more than 10 percent of the grizzly bears killed each year in Montana came from the Back Country. It was an ideal undeveloped recreation area because, as Forest Service personnel eventually would conclude, its fragile soils and shallow lakes would deteriorate under the impact of heavy recreational use. Moreover, its unspectacular scenery (by western Montana standards) did not make it the best place to locate scenic overlooks.[5]

Professor Behan attributed the origin of the controversy to the "telescoping" of events in the Lincoln area. In other parts of western Montana pressures to develop the national forests had been growing for nearly two decades, but in the Lincoln area they were compressed into a period of 3 short years. Before 1957 Lincoln was the epitome of the sleepy, isolated western town.

A one-lane dirt road was the only means of access to Lincoln from Ovando, 24 miles west, and the route was not better from the east side of the [Continental] Divide. Traffic between Missoula and Great Falls, the nearest population centers to the west and east, respectively, flowed in a devious routing through Helena, a detour of some 80 to 100 miles; the point is that no one ever drove through Lincoln. The road, when passable, was treacherous. But people could go to Lincoln, and just as the road provided ingress, so would it have provided egress for those who wished to leave. Most of the people in Lincoln, though, did not leave; they seemed to tolerate if not actually to prefer their isolated, quiet, idyllic winters, the superb hunting and fishing, and life the way it had been more or less since Lincoln began. In short, "progress" had effectively bypassed Lincoln and no one there seemed to mind too much, for it is a safe guess, provided the marginal roads out of town which were nevertheless roads, that the people in Lincoln were there because they chose to be, and they liked it.

Over the years, Lincoln became known as a base of operations for commercial guides and packers. It was a jumping-off point for entry into the Bob Marshall Wilderness by a low pass into the Danahar Valley. And not infrequently did the local outfitters take their "dudes" - mostly out-of-state people - into the Lincoln Back Country. By word and deed the regional and national reputations of the Lincoln outfitters grew, and so did that of the Back Country.[6]

Forest Service personnel were compatible parts of this setting. The Lincoln District Ranger and the Supervisor of the Helena National Forest had been on the job for nearly 20 years. They were representatives of the "custodial" era and were enthusiastic users of the Back Country. The Supervisor had formed a Back Country trail-riders group in Helena, and "both had shared the burden of long hours, low pay, relative obscurity, and no thanks that had been the accepted (and frequently the preferred) lot of Forest Service people for many years."[7]

As these two people prepared to retire in the late 1950's, plans were being laid in the Forest Service Regional Office in Missoula to develop the Back Country with a system of roads that would open it to timber harvesting and campground construction. The timber harvests were to pay for the roads and recreational developments that were the primary goals of the plan.

In 1957 Montana Route 20 had been completed linking Ovando, Lincoln, and Great Falls with a paved highway. In addition to truck traffic, the highway brought many automobile recreationists and potential mining development to the Lincoln area. A sawmill soon followed, adding 115 jobs to the local economy. Under these conditions of rapid change the situation was ripe for conflict between the Forest Service's desire to develop the Back Country and some local inhabitants who resented any disruption of their way of life. In fact, in later years the Forest Service occasionally would portray the struggle as one between progress and conservative resistance to change.[8]

According to Professor Behan:

On the one hand there was poised a vigorous and growing agency with a heritage of crusading hard work and the administrative toughness to resist local controversies. . .

On the other hand was anachronistic Lincoln, still a frontier town, an island in the riptide of postwar America. Here was represented the Rugged Individual, and a pioneering, roughing-it attitude that the people directed toward their wilderness of mountains, lakes and forests.[9]

By 1960 Lincoln residents had gotten wind of the Forest Service's development plans for the Back Country. In response to these rumors, three individuals, including a retired petroleum executive, William Meyger (who died in 1962), and a Forest Service campground foreman, Cecil Garland, formed the Lincoln Backcountry Protection Association. Although it was equipped with stationary and a letterhead, it had added only two new members by 1962 and was relatively inactive. The association's original goal was to delay development in the Back Country for about 10 years. They did not begin to lobby for wilderness designation until February 1964.[10]

Cecil Garland, a self-educated, "colorful character" from the Great Smoky Mountains of North Carolina, became the Association's president in 1962. He operated a hardware and sporting goods store in Lincoln and had worked four summers as a campground foreman for the Forest Service. He resigned when he realized that he could not pursue his goals from within the agency. Later he bitterly charged that the Lincoln Ranger District was over-staffed, that he had often been an unwilling "gold-brick" and that the Back Country was being developed to give idle Forest Service hands some useful work. While not accepting Garland's analysis of its motives, the regional office placed enough credence in his charges to direct the Helena Forest Supervisor to monitor more closely the district's administration.[11]

In 1962 Wallace Dresskill, assistant regional forester, gave his boss, Boyd Rasmussen, a portrait of the man who would be primarily responsible for the Scapegoat Wilderness Act of 1972.

Mr. Garland is an intelligent, energetic individual of 37. His formal education consists of elementary grades and four years at an agricultural vocational school in North Carolina. He had memorized his letter to you of the "wrongs" of the Lincoln District and practically repeated it verbatim. He started his discourse by describing the circumstances which led to the Declaration of Independence and compared it with the situation on the Lincoln District. He questioned the accuracy of the timber inventory, the economies of selling timber; the method of performing sale area betterment; the method of slash disposal; the practice of scaling private timber; the need for campgrounds; the judgment of the ranger in fire suppression; and the efficiency of the Ranger District administration...We will probably hear further from Mr. Garland.[12]

Tom Edwards, a former school teacher who had been an outfitter in Ovando for many years and was an early member of the association, was also an important figure in the effort to preserve the Lincoln-Scapegoat area. He travelled twice to Washington, D.C., to testify before congressional committees and in 1969 gave an eloquent personal testimony on behalf of the Lincoln-Scapegoat.

Into this land of spiritual strength I have been privileged to guide on horseback literally thousands of people - the old, many past 70, the young, the poor, the rich, the great and little people like myself. I have harvested a self-sustaining natural resource of the forest of vast importance. No one word will suffice to explain this resource, but let us call it the "hush" of the land. This hush is infinitely more valuable to me than money or my business...

The Forest Service proposed roads in this fragile land may satisfy the clamor of the masses but the hush of the land that the masses really seek will be crushed forever...

In consideration of these bills, from my point of view, it is unwise in the long run to be overly concerned about the outfitter for he is only a vehicle in the scheme of harvesting this wilderness resource. The group who really uses this resource and pays the bill is the public, the people the outfitter takes into the area. And what are these people buying? Is it fishing and hunting? Not for the most part. I would have gone broke years ago if this had been the case. As I said before, most come to this country to buy the "hush of the land"...

I know well the plan of the Forest Service. I've read it and discussed it with its authors and proponents. They sincerely feel that small islands of wilderness can be kept unsullied and undamaged. But how can we stand on that great Scapegoat Mountain looking down at its foot at bulldozers, trucks and cars at the heads of the Dry Fork, Cabin Creek, and Tobacco Valley, listening to the hideous noises of modern devices and trying to kid ourselves that we are enjoying the wilderness and partaking of its goodness...

This is the second time I've travelled across the Nation from Montana to represent a vast unseen audience who know this wilder-

ness because they have been there. . .I must keep faith with them. Around countless campfires year after year they have urged me to speak for them when and if the time ever came to save this Lincoln Scapegoat Back Country. . .

With all my being I urge you - don't let this majestic land down - don't let beer cans and the human filth that inevitably comes with a road lay waste to so priceless a heritage of our great nation.[13]

In March 1963 the Forest Service distributed its long-range plan for building roads into the Back Country. It also called for the exclusion of roads from 15,000 acres, which were to be placed in a Forest Service-designated Scenic Area. The plan had been approved by a majority vote of the Helena National Forest Advisory Council, an organization made up of a representative sample of users of the forest. Later some members claimed that they had approved the plan under the assumption that it would be put into effect over a 99-year period.[14]

Cecil Garland remembers his reaction to the possible implementation of the plan.

At that time a young Forest Service engineer quietly came into our store in Lincoln. . .and told me that the U.S.F.S. had abandoned a full survey of the road to the Lincoln Back Country and was now running only a "flag line" in their haste to build the road and then to quell the opposition. This young engineer in despair also told me that a bulldozer was sitting at the end of the road ready to drive into that country I had come to love above all else.

It was then that I knew that time was exceedingly short and in great desperation I went to the phone. . .

Finally I called Congressman Jim Battin and he answered the phone and I began to pour out my heart to him in a most pleading and earnest manner. Well, somehow he must have understood for he said he would help me and that he would send his aide up to visit us.

Congressman Battin then called Regional Forester Boyd Rasmussen on the phone and asked if he could have 10 days to see what was going on up at Lincoln. Mr. Rasmussen replied that Mr. Battin did not have ten days, that the bulldozer was ready to go.

Whereupon Congressman Battin told the Regional Forester that "By God, we had better have ten days." This incident is a classic example of democracy at work. Citizen goes to Representative; Representative goes to Bureaucrat. And at this time, I believe the tide turned in our favor.[15]

At the end of the month the Helena Forest Supervisor held a meeting at the Lincoln Lions' Club to discuss the plan. Some opposition was expressed, and the Forest Service decided to hold a full public meeting on the matter. On the evening of April 19, "some 300 people jammed into the small Community Hall in Lincoln." The Forest Service set the ground rules for the meeting - supporters and opponents were to alternate and there was to be no voice vote at the end of the meeting. Opponents of the development plan felt they had been "gagged," and a "near riot" took place. The level of bitterness over the plan began to increase dramatically. The association's membership rose to 50 and it soon received the backing of the Montana Wilderness Association, an organization Clif Merritt had helped form in 1956, and the Montana Fish and Game Department. Senator Lee Metcalf wrote the Forest Service asking it to delay development of the Back Country. During the next several months the Forest Service received no letters supporting its plan. The timber industry had expressed initial approval of the plans for timber harvesting but was heard from less and less as the controversy grew. According to Professor Behan: "Clearly, it seemed, the 'opposing interests' were the consensus [to delay development] on one hand and the Forest Service itself on the other."[16]

In June 1963 the Forest Service made a slight modification in the plan by eliminating part of a road, and in October the regional forester, Boyd Rasmussen, visited the area and came out strongly in support of the plan. Rasmussen's use of planning terminology to justify development of the Back Country illustrates why wilderness advocates were suspicious of, or at least ambivalent toward, the first modern forest planning legislation, the Multiple Use Sustained Yield Act of 1960.

In addition to these classified areas [wilderness and primitive areas], the National Forests of Montana include many thousands of acres of land that are primitive in character simply because planned development has not reached them. The Lincoln area is an example. These areas contain the bulk of future recreation development sites . . . Their timber is included in long-range timber management plans and the present allowable timber cut from the National Forests is based upon the eventual harvest of both the merchantable stands and the young growing stock which they contain.

These still undeveloped areas are composed of topographical units which are separable for discussion purposes but which must be tied into and treated as parts of larger integrated management units for intelligent development planning and the eventual use of their resources to serve the basic economy of Montana. Indiscriminate or patchwork "setting aside" of undeveloped land limits recreation potentials, decreases allowable timber harvests, and even more important may block or adversely affect the future management of adjoining areas.[17]

In late 1963 Robert Morgan became the new forest supervisor of the Helena National Forest. After looking over the situation, he decided to delay development until "absolutely necessary." (He was not personally for wilderness designation until several years later.) In a tactfully written memo in January 1964, Morgan told his superiors that although there was some passive support for the Forest Service's plan, "we will get no active support from the man on the street." He said the plan was "basically very sound" but that it was open to question on several points. He pointed out that the agency did not have a complete timber inventory of the area, that some timber of marginal quality had been sold, leaving an occasional "mess" behind, that neighboring national forests were not fully coordinating their plans with the Helena, and that the developed campgrounds around Lincoln were not being fully used. Morgan counseled the regional office that the Forest Service could probably win the Back Country battle if it were willing to go all out but that in the process it would pay a severe public relations price which might jeopardize some of its other programs in Montana. He concluded that:

The above approach appears as somewhat a compromise attitude. This is not good, and I realize the total effect may be a "drawing out" of the battle. On the other hand, we should maintain faith in managing according to actual resource needs and priorities. A stab at development before we are in a position with plans and finances to do a first class job could in the long run be worse than no development. I believe the need for developed recreation will assert itself undeniably, so that this argument will for the most part resolve itself in the interim. The on-the-ground job we do in management in the interim must also be sound enough to dispel apprehension.[18]

Morgan's "compromise attitude" was not well received in the regional office which wanted to begin road construction as soon as possible. Morgan and a succession of Lincoln District Rangers resisted that pressure. Over the next few years Morgan heard some rough words from his superiors, who undoubtedly questioned his loyalty and felt that he had caved in to local demands.[19] On the other hand, Clif Merritt felt that Morgan's temporary moratorium "was the wisest thing to do because the Forest Service began to look at the area more objectively and to talk to local citizens."[20] (Several years after the passage of the Scapegoat Wilderness bill, Morgan, with the congratulations of the regional office, received an award from an environmental group for his part in preserving the Lincoln-Scapegoat area).

When the Lincoln Back Country Protection Association met in February 1964, Cecil Garland convinced its members to support wilderness designation for the area because Bob Morgan could not commit the Forest Service to a 10-year moratorium. Garland also advocated that the wilderness be expanded to 200,000 acres to take in the Scapegoat Mountain region which adjoined the Bob Marshall Wilderness. Morgan reported that "it is clear that the group is determined to see that the area is preserved through whatever means possible. The group pretty well represents Western Montana conservation organizations."[21]

By this time The Wilderness Society in Washington was beginning to take note of the Back Country. In August 1964, just 3 weeks before the passage of the Wilderness Act, Harvey Broome, the president of the society, his wife, and Clif Merritt visited the area. According to Morgan, who escorted the trio into the Back Country: "Mr. Broome was obviously impressed with the area in question. I gathered that the society is definitely committed to support the request for the wilderness addition."[22]

Clif Merrit had camped in the Back Country as a boy and when he saw a road stake in his family's camping area he came to the sudden "violent" conclusion that they "would build a road there over my dead body." A few months later Merritt became The Wilderness Society's western regional representative in Denver and was a principal figure in the effort to get statutory protection for the area.[23]

In April 1965 Democratic Senators Lee Metcalf and Mike Mansfield introduced a bill to protect 75,000 acres of the Back Country under the Wilderness Act. Montana conservationists approached Republican Congressman Jim Battin and told him about the Metcalf-Mansfield bill and that there were more acres that could be included. Merrit remembers that "Big Jim had his big feet on a desk and when he heard this, they came down fast . . Jim saw this as an opportunity to leapfrog members of the other party." Battin introduced a bill calling for a 240,500-acre Lincoln-Scapegoat Wilderness. Metcalf and Mansfield, who, Merritt concedes, had not been fully informed about the situation, soon switched their support to the Battin bill.[24]

Some local citizens suggested that the Lincoln-Scapegoat be added to the Bob Marshall Wilderness but Clif Merritt successfully argued against that strategy. He contended that it would be bad "psychologically" because opponents could have replied that the 1-million-acre Bob Marshall was already big enough. The Bob Marshall rather than the Lincoln-Scapegoat then might have become the issue. Merritt, like Brock Evans, knew from experience that the environmentalists were strongest when they concentrated on individual areas.[25]

The Lincoln-Scapegoat bill was the first strictly citizen wilderness proposal made after the passage of the Wilderness Act. Since it did not involve the expansion of a primitive area, it was not explicitly covered by the study and review procedures of the Wilderness Act. The unique, potentially precedent-setting nature of the bill was one of the main reasons why its passage was delayed until 1972. The Forest Service leadership in Washington was concerned that its passage would unleash similar proposals at a time when its work force was committed to finishing on schedule the primitive area reviews mandated by the Wilderness Act.

In 1968 the Senate Interior Committee held hearings in Montana on the bill. Montana citizens and their four congressional representatives strongly supported it and little opposition was expressed. A retired forest entomologist and disgruntled Wilderness Society member expressed a minority view on the need for some developed recreational facilities in the Back Country.

Specifically, I am concerned that I shall no longer be able to enjoy the wonders of wilderness areas the creation of which I strongly supported. Why? Because I am getting too old to backpack into them and I cannot afford the prices asked by most outfitters. Furthermore, as an occasional hiker on short wilderness jaunts, I am increasingly infuriated by having to shuffle mile after mile along trails churned to powder by pack animals and liberally dotted with the manure of these animals used by the more affluent members of the wilderness set.[26]

As Bob Morgan later recalled, the 1968 hearings were "disastrous" for the Forest Service. Pointing to severe erosion caused by road construction in an area near the Lincoln-Scapegoat, Senator Metcalf testily asked Morgan how the Forest Service "could justify that". Morgan could only reply that "I can't."[27]

Soon after the hearings the Forest Service published a new plan for a 500,000-acre area, which included the Lincoln-Scapegoat. The plan called for some land to be administratively protected as "backcountry" and for the construction of a 75-mile scenic Continental Divide Highway through the Lincoln-Scapegoat. Local environmentalists were not placated. They argued that the Forest Service should have studied only the 240,000 acres of the Battin bill and that a Continental Divide highway which would be open only a few months of the year was not necessary.[28]

The Forest Service was becoming frustrated over an issue which refused to go away. In early 1969 this frustration moved Regional Forester Neal Rahm to tell a meeting of the agency's leaders that a "backcountry" land category, intermediate between complete wilderness and developed campgrounds, was needed. His remarks were also the first indication that the regional office was bowing to the inevitability of wilderness designation for the Lincoln-Scapegoat.

We have lost control and leadership in the sphere of Wilderness philosophy. Why? The Forest Service originated the concept in 1920, and practically, has been standing still since 1937...Why should a sporting goods and hardware dealer [Cecil Garland] in Lincoln, Montana, designate the boundaries for the 240,000-acre Lincoln Back Country addition to the Bob Marshall?. . . If lines are to be drawn, we should be drawing them.

All of this is slight tribute to Forest Service leadership and control. We seem to be trapped in our preoccupation with re-classification of Primitive Areas.[29]

In March 1969, 1 month after Rahm's remarks, Chief Ed Cliff told the Senate Interior Committee that the Forest Service would take another look at the Lincoln-Scapegoat. Plans for development were now permanently on hold. Two years later the forest supervisors of the Helena, Lolo, and Lewis and Clark National Forests drafted a wilderness proposal which the regional office accepted.[30]

The Senate passed the Scapegoat wilderness bill in 1969 and sent it to the House, where it was accidentally referred to the Agriculture Committee rather than the Interior Committee, thus arousing the ire of Chairman Aspinall who may have suspected an attempt to circumvent him. When he finally received the bill, Aspinall delayed reporting out the bill because the U.S. Geological Survey had not conducted a mineral survey of the area as called for by the legislative history of the Wilderness Act. The Montana congressional delegation requested the U.S.G.S. to make a special study of the area. Its study was completed in 1971 and showed no significant signs of mineralization.

Cecil Garland recalls how Aspinall was persuaded to support the bill.

> I had just left the House Office Building and Congressman Wayne Aspinall, the all-powerful chairman of the Interior and Insular Affairs Committee. Congressman Aspinall had just told me that he would "kill" my bill . . .
>
> Senator Mike [Mansfield] listened quietly leaning back in his chair, his fingertips touching gently as he moved his hands together again and again. And then he said, "Ceace, you go back to Montana and tell the folks back there that we'll get the bill passed, that there'll be a wilderness there some day." And he went on to say, "Some day there will be something that Mr. Aspinall will want, and we'll be there."
>
> We shook hands and I walked with him to the Senate floor where a great fight was being waged over Vietnam. But I knew Mike would not forget.
>
> Later when Congressman Aspinall became fully committed to the passing of the bill, I asked him why he had decided to help us. His reply was, "Son, you've got one powerful Senator", and I knew who he meant. I knew Mike had not forgotten.[31]

In 1972 the Scapegoat Wilderness became the first de facto wilderness to enter the National Wilderness Preservation System.[32]

As mentioned earlier, the Forest Service opposed the Lincoln-Scapegoat proposal because it did not want to disrupt its timetable for primitive area reviews. The regional office was also concerned that if the Backcountry Association were successful there would be petitions for numerous other de facto wildernesses surrounding the Bob Marshall Wilderness. This controversy illustrates a political science generalization. Agency behavior that is adapted to certain situations may become inappropriate when applied automatically to a circumstance which appears similar but is actually quite different. For decades the Forest Service had tried to insulate itself from local demands on the national forests in order to carry out its mandate to protect them in the national interest. These pressures usually came from groups that wanted to use them in ways that could have been detrimental to their long-term well-being. Environmental organizations and many in the general public supported the Forest Service when it resisted these demands. In the case of the Lincoln-Scapegoat, local pressure was also applied - not to use the forest but to protect it completely (or "lock it up" in the vernacular of the anti-wilderness opposition). The Forest Service fought this demand in the same way that it would have fought demands to over-cut or overgraze the area. The difference was that here the Forest Service was operating without public support.

This conclusion, however, must be qualified. A strongly professional organization, such as the Forest Service, is open to internal debate. Without the dissenting voices of Bob Morgan and the Lincoln District Rangers who served under him, roads would have been built in the Lincoln-Scapegoat long before the Scapegoat Wilderness Act of 1972.

Valdez Canyon, Pecos Wilderness Area
Sante Fe National Forest, New Mexico

(Above) The Red River Valley and Wheeler Peak
Carson National Forest

(Right) Gila Wilderness Area, Gila National Forest,
New Mexico

Bridger Wilderness Area
Bridger-Teton National Forest, Wyoming

Monongahela National Forest, West Virginia

Glacier Peak Wilderness
Mt. Baker National Forest, Washington

Three Sisters Wilderness Area
Willamette National Forest, Oregon

Shining Rock Wilderness Area
Pisgah National Forest, North Carolina

Wheeler Peak Wilderness Area
Carson National Forest, New Mexico

Chattahooche National Forest, Georgia

Spanish Peaks Wilderness Area
Gallatin National Forest, Montana

Sawtooth Wilderness Area
Challis National Forest, Idaho

Welcome Lake, Idaho Wilderness Area
Salmon National Forest, Idaho

Eagles Nest Wilderness Area
Arapaho National Forest, Colorado

Trappers Lake, Flattop Wilderness Area
White River National Forest, Colorado

Rainbow Lake, John Muir Wilderness Area
Sierra National Forest, California

Golden Trout Wilderness Area
Inyo National Forest, California

Flatside Pinnacle Observation Site
Ouachita National Forest, Arkansas

A Turn-of-the-Century Hunting Cabin
(thought to have been used by Theodore Roosevelt)
North Kaibab, Kaibab National Forest, Arizona

Tongass National Forest, Alaska

Kootznoowoo Wilderness
Tongass National Forest, Alaska

Tracy Arm-Fords Terror
Tongass National Forest, Alaska

Chapter V
Rare I and the 1975 Act on Eastern Areas

From 1972 to 1980 the issue of de facto wilderness was in the forefront of wilderness politics. Congress continued to pass individual wilderness bills based primarily on primitive area studies but they were overshadowed by several "omnibus" wilderness initiatives and the Forest Service's Roadless Area Review and Evaluation (RARE) studies.

For environmentalists de facto wildernesses existed in both the East and West. The Forest Service at first controlled the terms of the debate by maintaining that only a minimal amount of wilderness existed in the East. Thus environmentalists were forced to wage a separate struggle to convince Congress that de facto wildernesses in the East were "real" and that they should be protected under the authority of the Wilderness Act of 1964.

In 1967 the Forest Service manual directed regional foresters to review and report by 1969 on areas that might have wilderness potential. Later that deadline was moved ahead to 1972. Forest Service officials in the regional office in Missoula, Montana, were familiar with the section in the manual on roadless areas. The Lincoln-Scapegoat controversy had shown them the growing political importance of de facto wilderness. They wanted to begin a roadless study but were reluctant to proceed without national leadership from the Washington office.

In 1971 Dick Joy of the Missoula Recreation Staff spent a few weeks on detail in Washington. He told his Washington office counterparts of the regional office's interest in a roadless study and the need for uniform consideration and reporting from all regions. He believes this discussion helped begin the first RARE study.[1]

According to Chief Edward Cliff, RARE I was a response to a variety of forces that were impinging on the Forest Service:

Every time we made a move into a roadless area we ran into opposition which generally materialized in the form of a lawsuit or a wilderness proposal by a congressman. The principle of sovereign immunity had been breached in court cases in the 1960's. As a result, environmentalists started filing lawsuits and conservation law became a fast growing branch of the law. If a bill was pending, that effectively stopped any activity because we didn't want to aggravate Congress. We needed to draw some parameters around areas which we could develop and which we could preserve.[2]

Associate Chief John McGuire (promoted to Chief in 1972) recalls that he "sold" the idea to high-level departmental officials by arguing that the recently signed National Environmental Policy Act (NEPA) required environmental impact statements (EIS's) before roadless areas could be developed and that RARE I would constitute a national EIS. But he also remembers that neither he nor most of his Forest Service colleagues foresaw the full implications of NEPA. They thought the act called for relatively brief impact statements and not the massive, detailed tomes that the courts eventually required in most cases of federal actions affecting individual sites.[3]

In 1971 the Council on Environmental Quality tried to get the White House to issue an executive order halting development in roadless areas. The proposal got as far as John Ehrlichmann's office.[4] Some historians have suggested that fear of this order prompted the Forest Service to begin RARE I. Neither Cliff nor McGuire recall anything about it, although documents indicate that the agency's Legislative Affairs staff was concerned about it.[5] The evidence indicates that the decision to undertake RARE I was an internal one based on the general uncertainty over the management of de facto areas.

Between the fall of 1971 and the summer of 1972 the Forest Service inventoried and studied 1,449 roadless areas containing 55.9 million acres. With the exception of two areas in the East and one in Puerto Rico, all of them were west of the 100th meridian, which was a source of much disappointment to environmentalists. The Forest Service held 300 meetings and received more than 50,000 written and oral comments, which, at that time, made RARE I the most extensive public involvement effort ever undertaken by the federal government. The agency selected a list of 274 areas which would be protected while undergoing further study (the New Wilderness Study Areas) consisting of

12.3 million acres, of which 4.4 million had previously been committed to study either by the service or Congress.[6]

The Forest Service's hope that RARE I would settle the problem of roadless areas was dashed when the Sierra Club sued it for attempting to log an area that had been found unsuitable for further wilderness study. The Wilderness Society, which had committed most of its resources to the lawsuit over the Alaskan oil pipeline, again declined to participate because its council members predicted the Sierra Club's suit would not succeed.[7]

Judge Conti issued a preliminary injunction in August 1972 but before the case went to trial, the Forest Service Chief issued instructions to comply with NEPA before developing any roadless area. (In most cases this meant writing an Environmental Impact Statement before developing an area.) Judge Conti then dismissed the Sierra Club suit. Brock Evans was ecstatic over this perceived victory.

But perhaps the most brilliant victory of all was the injunction just obtained by the Sierra Club in San Francisco against the U.S. Forest Service, forbidding any more logging and road building on de facto wilderness lands on all National Forests until the Forest Service has complied with the provisions of the National Environmental Policy Act. Here is another great victory and I think the consequences and reverberations will be with us for years to come.

The most important thing, I think, is that for the very first time in years we are on an equal footing with the timber industry and the Forest Service with regard to our wilderness resource. The de facto wilderness areas no longer belong to them first as they have thought for so long.[8]

The Sierra Club and The Wilderness Society unleashed some of their harshest rhetoric on RARE I, charging that it had been done too quickly, that areas had been split arbitrarily into smaller units to lower their wilderness quality rating, and that the entire study was biased toward commodity values.[9] The Forest Service later conceded that RARE I was flawed but with the passage of time and cooling of emotions, environmentalists acknowledged its importance. According to Clifton Merritt:

I'm satisfied that at that time there was no full awareness of de facto areas. Finally RARE I and RARE II divulged a lot of those areas. Some of us were partially aware of them. Perhaps some of us were more aware than people in the Forest Service but still we had only had partial knowledge until those inventories.[10]

At the same time as the Forest Service was preparing for RARE I, its employees were discussing alternatives for managing undeveloped areas in the East. Public pressure for the designation of wilderness areas in the East had developed slowly after the passage of the Wilderness Act. In 1966 the West Virginia Highland Conservancy was formed by Rupert Cutler of The Wilderness Society and others with the goal of protecting areas on the Monongahela National Forest.[11] West Virginians' interest in wilderness appear, in part, to have been piqued by their opposition to the large clear cuts that had become common on the forest during the 1960's. The conservancy did not develop a consistent policy on wilderness until 1970.[12] Before that some of its members variously called for the protection of areas under the Wilderness Act, interim protection until these were restored as wilderness, or protection under new legislation.

The frustration and uncertainty that eastern wilderness advocates were feeling at the end of the decade comes across in a 1969 letter from George Langford of the conservancy's wilderness committee to Rupert Cutler.

A question of philosophy comes to mind . . .you represent an awful lot of apparently anonymous people who have not yet come forth in the name of The Wilderness Society. Where are they? Is Wilderness that foreign to the Easterner? Is the West Virginia Highland Conservancy an unwitting tool of The Wilderness Society? Are we the only people who care for Wilderness in the East?

Here we are asking to be set aside what was once wilderness but which has been exploited but which can again be wilderness (though probably not the same kind). These considerations enter to a much lesser degree in the West. Don't we need help on a grand scale to cope with this?[13]

In 1970 members of the West Virginia congressional delegation introduced several wilderness bills, as did members of the Alabama delegation for the Sipsey area on the Bankhead National Forest. These bills did not get out of committee, but they showed the Forest Service that public interest in eastern wilderness was mounting. The agency anticipated that it would have difficulties promoting alternatives to Wilderness Act protection because the Park Service had proposed 75,000 acres of wilderness in the Shenandoah National Park in Virginia, as well as parts of several wildlife refuges, which had been logged and burned over like most of the forests in the East.[14] The Forest Service realized that the public would not be satisfied with assurances of administrative protection and that if the Service did nothing, public "impatience may ultimately preempt Forest Service leadership in this area." In the summer of 1971 the Regional Foresters in Milwaukee and Atlanta proposed a "Wildwood Heritage System" that was to be distinct from the National Wilderness Preservation System. Forest Service Chief Max Peterson, who in 1971 was a deputy regional forester in Atlanta, remembers that the idea originated among agency officials in the South and East and that in the beginning the Washington office headquarters was not very involved.[15] The title was soon changed to "Wild Areas". According to agency policy analysts:

Wild Areas are distinct from Wilderness Areas because they are primarily for recreation enjoyment. Wildernesses are not primarily recreation areas, but are established primarily to ensure an enduring resource of wilderness for the nation as a whole. Wild areas would not qualify under the Wilderness Act. To include them would dilute the significance of the entire Wilderness System. Grazing and mining would be prohibited in Wild Areas whereas established uses are permitted in Wilderness Areas. Primitive recreation facilities and some hardening to protect the environment would be permitted in Wild Areas.

The Regional Foresters recommended that a "system" be established through broad enabling legislation. Individual areas could then be classified and included in the system after detailed study. The Chief feels that announcing such a system would force us to defend inclusion of the areas.[16]

Commercial timber harvesting was to be prohibited, but trees could be cut to improve wildlife habitat and recreation sites. The Forest Service also wanted to have the flexibility to do some "prescribed burning" in some potential "wild areas" in order to clear out undergrowth that made them virtually impassable. The ban on mining and grazing, which were permitted under the Wilderness Act, was consistent with the agency's penchant for pure wilderness. It may also have been incorporated as a gesture to please some environmentalists.

The public first became aware of Forest Service plans when Associate Chief John McGuire spoke before the Sierra Club's Biennial Wilderness Conference in Washington, D.C., on September 24, 1971.

The areas with wilderness characteristics as defined in the Wilderness Act are virtually all in the West. But we recognize that there is a pressing need in the East and South for providing primitive outdoor recreation opportunities and maintaining wild land values which do not meet the Act's criteria for wilderness ... We will solicit the broadest possible public participation in the process of creating and managing a system which would become a heritage of wild area values for future generations.[17]

McGuire remembers that he was subjected to some verbal buffeting at this conference from the young and somewhat unruly crowd. Doug Scott recalls that Ernie Dickerman ran up to him after the speech exclaiming: "Did you hear what that man said?"[18] Most professional environmentalists now consider this speech as the beginning of the general campaign for eastern wilderness.

During the next several months Forest Service officials met with conservation organizations, gaining the support of the American Forestry Association and of Joseph Penfold, conservation director for the Izaak Walton League. Timber industry officials were not enthusiastic about the idea, which they thought was just another wilderness proposal in a new guise. A political appointee with the U.S. Post Office Department suggested to the White House that the President support the wild areas proposal in his annual environmental message. Environmentalists also must have pressed their case, for when President Nixon delivered his message on February 8, 1972, it was a compromise between wilderness protection and the recreational emphasis of the wild areas concept. He urged Congress to make a great effort "to see that wilderness recreation values are preserved to the maximum extent possible in the regions where most of our public live."[19] Soon thereafter The Wilderness Society's lobbyists (Doug Scott, Harry Crandell, and Ernie Dickerman) persuaded White House officials to support the concept of eastern wilderness. The campaign received a boost when President Nixon came out in favor of eastern wilderness legislation in his Environmental Message to Congress of February 15, 1973.[20]

In July 1972 the Senate Agriculture Committee held hearings on several bills to create wild areas in the East. Two months later the Senate passed S.3973, the National Forest Wild Areas bill. Section 102 of the bill distinguished between the wild areas in the East which could be restored to a primitive condition and wildernesses in the West which had to be basically unspoiled by human contact. According to historian Ronald Strickland, S. 3973 "was first and foremost a statutory expression of the 'purity' doctrine."[21]

Behind the expression of the purity doctrine was the fear that if logged-over areas in the East were admitted into the Wilderness System, there would be no way to exclude logged-over areas in the West. The specter of numerous DuNoir controversies loomed before Forest Service policymakers. The Wilderness Society, on the other hand, believed that if the Forest Service were to succeed in establishing two systems and two sets of allocation criteria, it would be able to exclude all but the purest western forest land from the Wilderness System.[22]

The Wilderness Society's staff were completely caught off guard when S. 3973 passed the Senate. Emergency meetings were held and Doug Scott typed the draft of an alternative eastern area wilderness bill with a speed which was becoming one of his trademarks within the wilderness movement. Several months earlier he had met Senator James Buckley of New York who had told him that he had been a long-time member of The Wilderness Society. Scott pencilled in the names of Buckley and Frank Church as sponsors of the draft bill. Senator Church was out of town but Scott encountered an aide to Interior Chairman Senator Henry Jackson who said that Jackson might want to co-sponsor the bill because of the issue of committee jurisdiction over eastern wildernesses.[23] Both Jackson and Buckley agreed to sponsor S. 3792, which was re-introduced in January 1973 as S. 316. Ernie Dickerman has recalled the early stages in the development of S. 316.

Wherefore with the opening of the 93rd Congress citizen wilderness proponents living in the East were pounding on the doors of the Interior Committee. As I recall, close to 100 proposals located in maybe 15 states were offered. When Senator Haskell convened a public hearing in Washington on this assortment of proposals, over 150 individuals from most of the New England, southeastern and midwestern states were on hand, most of them ready and eager to testify in person. Senator Haskell's problem was how to accommodate all these witnesses and to avoid endless repetition of the same arguments. The hearing lasted most of the day. Meanwhile these numerous citizens, when not testifying, were lobbying all over Capital Hill. Ultimately what emerged from the Subcommittee on Public Lands, rather than a number of separate state bills, was a single omnibus bill containing all proposals which the Committee was to approve. This omnibus bill of course made possible a single, coordinated drive for its passage.[24]

The Senate Interior Committee had handled all previous wilderness bills dealing with western forests and thus claimed jurisdiction over eastern wilderness bills. The Agriculture Committee had jurisdiction over the national forests in the East because their lands had been purchased from private owners rather than reserved from the public domain like those in the West. Therefore, its members also asserted jurisdiction over eastern wilderness bills. This institutional rivalry was paralleled by a personal coolness between the chairmen, Senator Herman Talmadge of Georgia and Jackson, who communicated only through their aides.[25]

The Wilderness Society was distressed as much by the possibility that the Agriculture Committee might control eastern wilderness bills as it was by the possibility of a separate wild areas system. According to Doug Scott, The Wilderness Society "had a healthy perhaps obsessive respect for the Forest Service's influence with the Agriculture Committee."[26] That respect seems to have been justified because Chairman Talmadge professed to have more faith "in the management plans being drawn up by the Forest Service to provide true multiple use, . . .than on designations forced upon the Forest Service by strident interest groups."[27] The ranking Republican member of the committee, George Aiken of Vermont, was an "old friend" of the Forest Service and often shared his home remedies with John McGuire when McGuire visited his office on Saturday mornings.[28]

The biggest problem that environmentalists had to overcome was dissension within their own ranks. Many people were willing to accept (or at least did not oppose) the wild areas proposal because it seemed to promise that eastern areas would be protected under legislation that, in some respects, was even stronger than the Wilderness Act. The semantic struggle over wilderness versus wild lands seemed less important to them than the possibility that, to paraphrase Richard Costley, commercial lumbering and roads would be kept out of their favorite areas.

In addition, some conservationists had reasons for opposing the eastern wilderness concept. Some northeasterners had become skeptical about the wisdom of wilderness designations after observing the heavy use of the Great Gulf Wilderness in New Hampshire, while members of the Appalachian Mountain Club, who had good relations with the Forest Service, were afraid that their trail shelters would have to be removed from wilderness areas. Ernie Dickerman also believes that the New Englanders' traditional feeling of independence played a role in delaying the eastern wilderness movement. New Englanders were justifiably proud of their conservation record and some felt that wilderness legislation was an unnecessary imposition.[29] Senator Aiken's support of the wild areas concept may have reflected this attitude.

One of The Wilderness Society's most important opponents on this issue was Joseph Penfold, venerable conservation director of the Izaak Walton League, who had been one of the leaders in the passage of the Wilderness Act. Penfold had "endless credentials from which to speak"[30] and the prestige of the Izaak Walton League behind him when he warned in Bill Worf fashion of the dangers of designating eastern wildernesses.

> *It is argued by some that such areas reverting to wilderness should be eligible for inclusion in the wilderness system. They argue that the criteria and standards established by the Act are sufficiently flexible now to embrace such areas . . . Others, including the writer, feel just as strongly that lowering wilderness standards by amendment of the Act or by its more liberal interpretation, in the long run, can only threaten the integrity of all designated wilderness. Even if the line can be held rigorously against the invasion by commercial development - and pressures for such are unrelenting - there still remains the growing and nearly irresistible pressure of recreationists themselves who, with snowmobiles, outboard, ATV and other gadgetry, or sheer numbers lean their weight against every wilderness boundary."[31]*

George Alderson of the Friends of the Earth lobbied members of the Izaak Walton League to support eastern wilderness at their 1972 annual meeting in New York. The Indiana Chapter was the first to break ranks and soon thereafter the entire league came out in support of The Wilderness Society's position.[32]

Ironically, the society had the most difficulty with the Sierra Club, an organization with which the society's leaders always had been closest to philosophically and personally, since both shared common officers and routinely exchanged staff members.

Their conflict in this case arose largely from their differing organizational structures. The Wilderness Society is a centralized organization with a large Washington staff. It has a governing council which sets policy, and has dues-paying members, but has no local chapters. It works on the local level through cooperators and consultants who share the society's objectives on a particular issue but need not be members of the society.

Its officers and staff members felt they were the heirs of Howard Zahniser's wilderness legacy and were completely unwilling to compromise that legacy by creating two systems. Ernie Dickerman had known both Zahniser and the society's long-time president, Harvey Broome, and was convinced they had never meant to exclude eastern wilderness when they wrote the drafts of the wilderness bills.[33] In addition the society has had a large number of naturalists and ecologists as members and leaders, and the concept of a single, unified system was more compatible with their scientific background. In 1972 Ernie Dickerman explained to Hal Scott of the Florida Audubon Society that the

> *"definition of wilderness in the Wilderness Act was drawn by Easterners, i.e. by Howard Zahniser, then Executive Director of The Wilderness Society, who gained his formative experiences of wilderness in the Adirondacks, and by Harvey Broome, then president of The Wilderness Society, life-long resident of Knoxville, Tennessee, who hiked and camped for nearly fifty years in the Great Smoky Mountains and other areas in the southern Appalachians. None of us who helped draft that definition were about to exclude the wild lands of the Appalachian Range from qualifying for the protection of the Wilderness Act. . .*
>
> *. . .In our opinion it is preferable to put into the System an area which may contain some minor work of man than it is to reject the entire area or a significant portion of it in order to avoid such minor features. After all, the objective is to preserve wilderness, not seek reasons for rejecting its preservation. Once an area is in the System, however, we can expect to fight with maximum skill and diligence to prevent even the most minor sort of intrusion into a legally designated wilderness area."[34]*

The Sierra Club is larger than The Wilderness Society and its members are organized into local chapters which are overseen by regional field representatives. Many of them are hikers, backpackers, and mountain climbers. In general the club is more "user-oriented" than The Wilderness Society.[35] During 1972 and 1973 the club's Midwest and

Northeastern field representatives supported wild areas legislation as the best way to protect eastern areas given the Forest Service's opposition to wilderness legislation. Ted Snyder's report of a March 1972 meeting involving himself, Jonathan Ela, Francis Walcott of the Sierra Club, and Forest Service officials show the club's tentative attempts to reach a compromise with the agency.

> *Theodore A. Snyder opened in the morning by proposing some wilderness in the National Wilderness Preservation System now, to be followed by "Future Wilderness" Act, with no tinkering and automatic addition to Wilderness Act on meeting a definite standard. Forest Service countered even recovered lands don't meet definition because the Wilderness Act refers to helping land 're-tain' its primeval state.*
>
> *Ela opened in afternoon. Suggested laying Wilderness Act "on the shelf" and talk about wild areas; if we could not agree on wild areas, then to bring Wilderness Act "off the shelf" and talk about applying it to eastern areas. Theodore A. Snyder countered with proposition of the morning - some wilderness now, then other land into "future wilderness". Snyder listed wilderness now areas as Caney Creek, Ouachita National Forest; Headwaters of Buffalo River, above National River Boundary, Ozark National Forest; Cohuttas, Georgia; Bradwell Bay, Florida; Kilmer-Slickrock, North Carolina; Sipsey, Alabama; three areas in West Virginia impending bills; Penigewasett in White Mountains National Forest.*
>
> *Forest Service people said Sipsey had only 400 acres of virgin forest. I said they were wrong.*
>
> *Forest Service said of this list only one that might qualify was Kilmer-Slickrock...*
>
> *Discussion of necessity of transfer to Wilderness Act - if it's a rose by another name why worry? Ela asks why not put the Wilderness now on list into the new Act? Would not the same thing be accomplished by getting them into wild area protection, and perhaps more because of mining and grazing. Snyder says no; Wilderness would be quicker. Ela says we can get a good wild areas law just as fast.*
>
> *Ela proposes a Wild Areas Law as "Future Wilderness" described by Snyder, but without automatic turnover. Proposes it be structured like Wild and Scenic Rivers Act, with a list of instant Wild Areas, and a list of mandatory review areas, with further provision for review of other areas.*
>
> *General agreement this would only apply to East. Forest Service seemed in agreement; said this would create a class of land approaching characteristics they desired in wilderness.*
>
> *Discussion of why these areas could not eventually become wilderness because no one can say what would have been there if no human activity. Walcott and Snyder point out that even virgin forest has cycles; all you need to do is get the land onto a natural cycle.*
>
> *Ela, Walcott, and Snyder agree no response should be made until Board of Directors acts in May."*[36]

The Sierra Club's northeastern field representative, who was not present at the March meeting with the Forest Service, worked with Senator Aiken of the Agriculture Committee on his wild areas bills. According to Doug Scott, the Agriculture Committee's present co-jurisdiction over eastern wildernesses is, in part, a result of that relationship.[37]

Many of the club's eastern members also supported wild areas legislation as the best way to protect areas they were interested in, and were either unaware of or unconcerned with larger political issues. In early January 1973, Helen McGinnis, a Sierra Club activist and member of the West Virginia Highland Conservancy, listed four reasons why local club members supported the wild areas proposal.

> *1. Misinterpretation of the Wilderness Act of 1964 and the alleged superiority of wild areas legislation...*
>
> *2. The "pragmatic approach", which assumes that the Forest Service is too powerful to oppose, and that environmentalists must cooperate with it to get any areas protected.*
>
> *3. The feeling among some conservationists that the Forest*

Service is doing a good job of managing the national forest and if they prefer wild area to wilderness, well they're the experts. As a member of the West Virginia Highlands Conservancy, which has many disagreements with the Forest Service concerning the management of the Monongahela National Forest, and the Sierra Club, which so far has excellent relations with the staff of the Allegheny National Forest, I can understand both viewpoints to some extent.

4. The belief that eastern wilderness legislation should be supported, but that there is also a need for legislation to protect other areas not qualified as wilderness...[38]

Hoping to gain allies for its uncompromising position on eastern wilderness, in December 1972 the society invited a select group of environmentalists, including local Sierra Club leaders, to a meeting in Knoxville, Tennessee. There were "skeptics" in this group but all were chosen because they were considered to be potential converts. The Sierra Club's regional representatives, who were thought to be less persuadable, were not invited. Also present were two staff members from Senator Church's Public Lands Subcommittee who helped convince some of the doubters by supporting all of the statements made by the society's spokesmen. According to its organizers, Doug Scott and Ernie Dickerman, this meeting was the first step in overcoming Sierra Club resistance to eastern wilderness legislation.[39]

The Wilderness Society also formed a very successful "front" organization called "Citizens for Eastern Wilderness," the main purpose of which was to be a face-saving device for Sierra Club members who wanted to discreetly align themselves with the society.[40] In March 1973 George Alderson culminated this campaign with an influential article that strongly underscored the alleged political dangers of wild areas legislation.

The wild areas system would serve the Forest Service cause well. It sets two Senate committees to fighting over specific wilderness area proposals; it confirms the Forest Service "purity" argument; it fragments the wilderness movement into regional factions with less influence, instead of a unified national movement; it puts the citizen environmentalists at a disadvantage in unfriendly committees of Congress. And it gives the Forest Service new hope for stopping the citizens' wilderness proposals in every western state, from California to Montana.

The Forest Service's objective on Capitol Hill is evidently to get its eastern lands firmly away from the Interior Committee, where citizens have a great deal of influence, and let the Agriculture Committee do the dirty work of turning down all the wilderness or wild area proposals.[41]

The Sierra Club eventually endorsed the society's position but not before there had been "difficult" meetings and some bad feelings between officials of the two organizations. For instance Oregon native, Doug Scott, at first decided not to apply for the important position of northwest field representative of the Sierra Club because of disenchantment over the club's approach to eastern wilderness. He was later offered the position, which he accepted.[42]

In 1973 the Senate Interior Committee (later becoming the Energy and Natural Resources Committee) and Agricultural Committee agreed to share jurisdiction over eastern wilderness legislation with the understanding that eastern areas would be protected under the Wilderness Act of 1964. By early 1973 the Forest Service had dropped the idea of two systems. In September the Department of Agriculture proposed that 16 eastern areas be made instant wildernesses and that another 37 be studied for possible inclusion in the Wilderness System. But this proposal also contained a section which would have amended the Wilderness Act of 1964 to read that "only within those national forest system units east of the one hundredth meridian the Secretary of Agriculture may consider for review areas where man and his own works have once significantly affected the landscape..." The environmentalists defeated this attempt to establish separate criteria for eastern wilderness. Instead, the Agriculture Committee reported out a bill that extended a ban on mining and grazing previously limited to the East, to the entire Wilderness System. Chairman Jackson and western members of the Interior Committee strongly objected to this provision, which they and the environmentalists knew could prove fatal to the possibility of enlarging the Wilderness System in the West. The extension of the ban on mining and grazing to the West was voted down on the Senate floor and the bill was referred to the House.[43]

The death of John Saylor in 1973 and Washington's preoccupation with the Watergate scandal delayed House passage of the bill for 8 months. House Public Lands Subcommittee Chairman, John Melcher, required that all areas included in the bill be supported in writing by the representative in whose districts they were located. (Ernie Dickerman saw this as a departure from the usual protocol requiring only oral assent.[44]) This procedure had the effect of halving the number of acres that were to go immediately into the system or that the Forest Service was to study for possible future designation. In December 1974 the House passed the bill. It differed from the Wilderness Act only in allowing the federal government to condemn private land under defined conditions when it lay within the boundaries of the eastern wildernesses in the bill. It gave the secretary of agriculture the authority to condemn land whenever "he finds such use to be incompatible with the management of such area as wilderness and the owner or owners manifest unwillingness, and subsequently fail, to promptly discontinue such incompatible use." The Wilderness Society was disappointed that it had lost potential wilderness acreage during the bill's passage through the House but rather than wait for a new Congress to convene, Dickerman advised Senator Jackson that the society would accept the House bill.[45] The bill, which contained 15 wildernesses and 17 wilderness study areas, was signed by President Ford on January 3, 1975. It has erroneously been called the "Eastern Wilderness Act"; however, it has no title and is one of many acts designating units of the Wilderness System.

Chapter VI
The Endangered American Wilderness Act and RARE II

Because the 1975 act for eastern areas designated wildernesses that had once felt the heavy impact of "man and his own works," one former member of the Forest Service's Legislative Affairs Staff dubbed it the crossing of the "last promontory of purity."[1] Ernie Dickerman accepted the final House bill, despite the fact that it did not contain as many acres as he had wished, because he realized its passage would be the most significant victory for The Wilderness Society since the passage of The Wilderness Act of 1964.[2] But if the last promontory had been crossed in January 1975, there still remained islands of purity which became the environmentalists' next target.

RARE I had been strongly criticized by the wilderness organizations for failing to select as New Wilderness Study Areas several potential wildernesses located near major population centers. Some of these, such as the French Pete area in the Willamette National Forest in Oregon, were also marked by strong conflicts between the timber industry and environmentalists. French Pete assumed great symbolic importance for all the interest groups involved. Its historical interest is further enhanced because Doug Scott helped conceive the idea for the omnibus Endangered American Wilderness bill partly to preserve this area, which he had learned about while growing up in Oregon.[3]

In 1953 the secretary of agriculture eliminated 53,000 lowland acres from the Three Sisters Primitive Area. In 1954 the regional forester in Portland explained to a Sierra Club representative that the area did not contain any unique flora and that harvesting some of its 1.5 billion board feet of timber would help to prevent the closing of more mills in the area. The Willamette Forest Supervisor pointed out that 60 percent of the sawtimber in Lane County was controlled by the Forest Service and that competition for logs was great, which placed "tremendous responsiblities" on the agency to insure the viability of the local lumber industry. Local environmentalists were not persuaded by these arguments and appealed the decision to the secretary of agriculture, who reaffirmed the 1953 decision.[4] By the late 1960's the 30,000-acre French Pete Creek area was all that remained of that former unroaded and unlogged area.

In 1968 the Willamette Forest Supervisor, David Gibney, announced a timber sale in French Pete, a popular hiking and camping area about an hour's drive from the towns of Eugene and Springfield. Local environmentalists responded by forming a "Save French Pete Committee" and petitioning the Forest Service to withdraw the announcement. Gibney assembled a citizen's advisory committee of 23 people who represented various users of the forest. In March 1969 they voted 18 to 5 to support Gibney's plan to selectively log the area. Gibney concluded that the evidence "shows that demands and needs for all resources have increased proportionately since 1957, whereas the forest land base has remained almost static. Therefore, the need for a review and recommendation of the decision of the secretary of agriculture in 1957 cannot be substantiated."[5]

In the meantime the Save French Pete Committee had been organized as the result of the exhortations of the Sierra Club's new northwest regional representative, Brock Evans, who had motivated a previously dispirited group of local environmentalists. Evans had obtained a large map of Oregon's forests, which showed that French Pete was one of only three valleys, 10 miles long or more in the State, which had not been logged. That discovery mobilized a "tired band of warriors for one more effort" and soon became the committee's principal rallying cry.[6] In November 1969, the Forest Service's predicament was dramatized when 1,500 protesters gathered outside the Eugene Federal Building to hear several speakers attack its decision.

National wilderness leaders saw French Pete as the most important environmental issue in the Pacific Northwest and as a major test of strength with the region's powerful timber industry.[7] Michael McCloskey asserted that if the environmentalists could win on French Pete, they could win anywhere.[8] Although the Forest Service had proposed to sell only 3 million board feet of timber, industry representatives saw the struggle in equally apocalyptic terms, predicting "further incursions in the areas where there is much commercial timber" if the environmentalists prevailed.[9]

The conflict between environmentalists and the timber industry was paralleled by a split within Oregon's congressional delegation. Bob Packwood, the newly-elected republican senator, supported the environmentalists and spoke of French Pete as the environmental counterpart to the debate which was then raging over the deployment of an antiballistic missile system. He was opposed by veteran Senator Mark Hatfield and Congressman John Dellenback, who represented the district in which French Pete lay.[10]

In 1969 Packwood introduced the "French Pete Intermediate Recreation Area" bill, which permitted the development of some recreation facilities and the harvesting of dead and down timber. For several years environmentalists did not attempt to introduce a wilderness bill for French Pete, fearing that it would be impossible to pass such a bill in the face of the timber industry's determined opposition. The "purification" of the neighboring Mount Jefferson Wilderness frightened some people, who did not want a similar fate to befall French Pete.[11] The Wilderness Society called for wilderness designation but their arguments had little immediate effect because Oregon was in the Sierra Club's informal sphere of influence. Environmentalists, like some of their counterparts in the East, were looking for a practical way to exclude logging and roads from French Pete and did not want to sacrifice it "on the altar of Wilderness nationwide."[12] Michael McCloskey, executive director of the Sierra Club, lent his support in 1972 when he said that French Pete could be a prototype for areas "between full wilderness and roadside recreation facilities."[13]

Packwood's bills were blocked by Senator Hatfield, who was a member of the Senate Interior Committee. In 1972 Hatfield made a gesture to the environmentalists by asking the Forest Service to delay its plans until it could study the feasibility of using helicopters to log the area. French Pete became one of the issues in his 1972 re-election campaign against former Senator Wayne Morse, who claimed that Hatfield had received campaign contributions from loggers near French Pete.[14] His victory over Morse demonstrated Hatfield's continuing popularity in the State, and confirmed that Oregonians were split over French Pete, with a substantial minority favoring some development of the area.

From 1969 to 1976 the Forest Service's development plans were repeatedly postponed by agency and departmental officials, who gradually realized that they would probably never be able to implement those plans. But a continuing political deadlock frustrated hopes for a permanent solution to the problem. In 1973 local environmentalists and Senator Packwood dropped their Intermediate Recreation bill in favor of adding French Pete to the Three Sisters.[15] It was not until the 45,000-acre area was "embedded" in the Endangered American Wilderness Act of 1978, where it was less exposed to attack, that the political impasse was finally broken. (Environmentalists are now fond of saying that the longer the Forest Service delayed, the bigger the area became.[16]) A collective sigh of relief went up from many Forest Service officials who had been involved with French Pete over the years and who believed that the 1957 decision had tied their hands. Former Willamette National Forest Supervisor, Zane Grey Smith, saw French Pete in much the same light as did Doug Scott. When Assistant Secretary Cutler decided to support the Endangered bill, Smith looked at the clock and said to Cutler that "on this date and at this time French Pete was finally solved."[17]

As the French Pete controversy quieted down in the mid-1970's, the environmentalists' main theater of operations in the West shifted to the vast expanses of undeveloped forest land in the Rocky Mountains. The Forest Service had promised the Court in the Sierra Club v. Butz suit that it would comply with NEPA before developing any roadless areas inventoried in RARE I. An EIS was prepared for a "unit," which was usually about the size of, but not necessarily coterminus with, a ranger district.

David Pavelchek, Doug Scott's employee in the Sierra Club's Seattle office, had an "extraordinary " ability to quickly analyze the technical prose of an EIS and memorize the information it contained. He was set to work dissecting the Forest Service's unit plans.[18]

On the Kootenai National Forest, Pavelchek discovered that a unit plan had included several roadless areas and that development decisions were based on an overall evaluation of the unit. The Sierra Club appealed the decision to Chief John McGuire, claiming that one of the roadless areas, Mount Henry, should have been evaluated separately. McGuire upheld the Sierra Club's appeal.[19] (In 1977 Mount Henry was one of the areas included in the Montana Wilderness Study Act but was "released" for nonwilderness use in the Lee Metcalf Wilderness Act of 1983.)

The converse of Mount Henry was the Gospel Hump area on the Nez Perce National Forest in Idaho. Here Pavelchek found that the Forest Service had divided the nearly 300,000 acres of roadless area into nine separate units. The Sierra Club maintained that the wilderness value of Gospel Hump would have been higher if the Forest Service had studied it as a single unit. Doug Scott recalls the crestfallen expression on McGuire's face when he was told of this situation. Again, the Forest Service chief upheld the Sierra Club's appeal.[20] (Gospel Hump was included in the Endangered American Wilderness Act.)

Pavelchek's work shook the Forest Service's confidence in the unit planning process and compelled it to write stricter guidelines for appeals so that its planning efforts could not be stopped completely. It also convinced Doug Scott that a "phase change" was needed before the

Forest Service adjusted to the Sierra Club's tactics.[21]

While talking with his friends and Sierra Club colleagues, Chuck Clusen and John McComb, in January 1976, Scott recalled the omnibus de facto wilderness bill that John Saylor had introduced in 1970. The three Sierra Club staff members agreed that the time was ripe for a new omnibus bill.[22] Thus was born the Endangered American Wilderness bill.

The Endangered bill contained several areas scattered throughout the West (the act contained 17), which the Forest Service had not recommended for wilderness study in RARE I, primarily because they did not conform with various aspects of the agency's definition of wilderness purity. One of the main issues concerned the so-called "sights and sounds" doctrine that had been used to exclude areas that were close to major urban centers. Sandia Mountain, Lone Peak, and Pusch Ridge, which overlooked Albuquerque, and Salt Lake City and Tucson respectively, were chosen to illustrate the application of that "doctrine". The final committee reports on the bill directed the Forest Service to dispense with this doctrine, arguing that the accessibility of such areas actually enhanced their values as wilderness.[23]

The campaign was guided by Doug Scott and Chuck Clausen, who worked with local environmentalists in selecting the areas to be included in the bill. Like the Wilderness Act of 1964 and the 1975 Act on eastern areas, the Endangered bill was coordinated by staff members of the national wilderness organizations. It was the most programmatic of the three omnibus bills. Most of the areas chosen for inclusion in the bill had been fought over for many years by environmentalists and their opponents. The environmentalists wanted to get them into the Wilderness System but also wanted to make them symbols of supposed defects in the Forest Service's planning process.[24] Michael McCloskey explained the purposes of the bill to George Davis, then executive director of The Wilderness Society.

To get a number of high quality, seriously threatened national forest "de facto" wilderness areas in the Wilderness System.

To educate the Congress to the underlying problems of land use planning on the national forests, particularly the inadequate consideration of wilderness values of these roadless areas, and

To build a nationwide force of grassroots support for wilderness, using this bill as the vehicle, and thus reminding every Senator and Congressman of the broad support we can marshall for wilderness issues, particularly from Eastern and more urbanized districts.[25]

Between the first introduction of the bill in 1976 by Senator Frank Church and Representative Morris Udall and its enactment in January 1978, there were some changes in the areas included. For the most part, the Sierra Club coordinators chose areas that they judged were not so controversial that they would block final passage of the bill. According to Doug Scott: "Each area was ready to go but didn't look it - that was the stroke of genius about it."[26]

During most of 1976, congressional mail had been running against the Endangered bill, but by the fall the Sierra Club's campaign started to gain momentum, and Congress began to receive more mail and delegations of citizens favoring the bill. Senator Church and Representative Udall were presidential candidates. Their advocacy of the bill also convinced Jimmy Carter to support it. Carter's election in November 1976 considerably increased the chances that the bill would pass. During the presidential campaign, Carter's staff had promised the environmentalists strong support and in one case had even gone beyond what they had asked for, by offering to place a development moratorium on all roadless areas. George Davis bemoaned what he thought was a "bad" National Forest Management Act, which Congress had passed earlier in the year to extricate the Forest Service from a court-imposed injunction on timber harvesting. He told Carter officials that he feared a similar "backlash" from the timber industry if a moratorium were placed on all roadless areas.[27] The proposal was soon dropped.

The passage of the bill was virtually guaranteed when Morris Udall became the new chairman of the House Interior Committee. Upon receiving news that Congressman "Bizz" Johnson would take the Public Works Committee and that he would then chair Interior, the 6'5" Udall hugged the much smaller Chuck Clusen, exulting "now let's pass that Endangered bill."[28]

The Endangered bill was still being debated when Rupert Cutler became the new assistant secretary of agriculture for conservation, research, and education. Previously the Forest Service had not taken a position on the bill but Cutler persuaded agency officals to support it. Cutler had some reservations, however, because although he was urging the Forest Service to liberalize its definition of wilderness, he did not have any criteria by which to judge whether the "endangered" areas were any more worthy of wilderness designation than the hundreds of other roadless areas in the National Forest System.[29]

Cutler had resigned from his position as assistant executive director of The Wilderness Society in 1969 to pursue a Ph.D. degree in resource economics and law at Michigan State University. After he completed his dissertation, dealing with several citizen lawsuits against the Forest Service (including the Parker Case) in 1972, he joined the faculty of the Department of Resources Development at Michigan State University. He maintained his affiliation with the Sierra Club and The Wilderness Society, but also was appointed by the Governor to several resource commissions in Michigan, where he acquired a "better" understanding of all sides in environmental disputes.[30]

After Carter's election, Doug Scott and Larry Williams of the Oregon Environmental Council discussed possible candidates for the position of assistant secretary of agriculture for conservation. They settled on Cutler, who had brought Scott into The Wilderness Society 8 years earlier. They considered him to be the best choice because of his background in the environmental movement and his solid academic credentials in natural resource management. Scott called Cutler and suggested that he apply for the job, as did Brock Evans. Cutler went to Washingtron, D.C., and "lobbied for the job on his own" without further help from the environmental organizations.[31] Cutler apparently was the only person considered by Secretary Bob Bergland, who lectured Cutler on the congressional committee members he should cultivate and then gave him complete independence to administer the agencies under his charge after he had been confirmed.[32]

The prospect of Cutler as assistant secretary distressed some timber industry officials who remembered his tenure with The Wilderness Society. He had already sold his home in Michigan and had moved to Washington. Now he faced the possibility that the timber industry might try to block his appointment. In order to quiet their fears, he and Secretary Bergland met with a group of industry officials at a hotel near O'Hare Airport in Chicago, where they were told of the industry's difficulties in gaining access to national forest timber. Industry spokesmen had prepared a map of Idaho showing the areas which were in wilderness, primitive status, wilderness study, and in litigation. Only a small percentage, highlighted in green, was shown as open to harvesting. Cutler expressed sympathy for the industries' difficulties in making investment decisions in this situation and said that he would look into the matter and make a review.

R. Max Peterson (then deputy chief for programs and legislation) recalls that Cutler returned from Chicago saying that everybody had been impressed by the "uncertainty" over the wilderness situation and that he had made a commitment to accelerate the examination of roadless areas as a prelude to land management planning for all of the national forests. He directed the Forest Service to do a "better" job of inventorying roadless areas than it had in RARE I. Unlike RARE I, which only selected areas for further study as wilderness, he wanted as much as possible to resolve the "uncertainty" by recommending some areas for wilderness designation and "releasing" others. Agency officials asked the department's lawyers about the legality of a roadless study and were told that they were in "uncharted waters." There was nothing in either the Forest Service regulations or case law concerning the kind of large, programmatic EIS that Cutler envisioned. All previous NEPA court cases had involved individual sites. The agency also did a probability study which concluded that there was a very slim chance that the study would completely resolve the wilderness issue. It predicted that a programmatic EIS would be tested in court and that roadless areas would gradually be allocated in statewide bills. (Both of these predictions came to pass within the next few years.) Agency officials were also aware that Congress had not acted on several primitive area recommendations, all of which had been submitted by 1974, and therefore were not optimistic that the legislators would act quickly on the results of a nationwide wilderness study. In addition, they were concerned that a roadless study would place a heavy burden on field personnel who also had to gather and analyze data for the Resource Planning Act Program in 1980. Thus, the Forest Service leadership was not enthusiastic about Cutler's proposal and would not have done it "if left to its own devices."[33] Once in office and confronted with the Endangered bill, he decided to go ahead with this plan, announcing it at a congressional hearing.[34]

The environmentalists were totally unprepared for Cutler's announcement. Weeks earlier Doug Scott and other environmentalist leaders sent angry telegrams to Cutler, their friend and former colleague, concerning his testimony on the Oregon Omnibus Wilderness bill, sponsored by Senator Hatfield. (The areas in the bill were later included in the Endangered American Wilderness Act of 1978.) Scott speculates that Cutler reacted to these expressions of outrage and the timber industry's problems by conceiving RARE II. Skeptics in the wilderness movement suspected the study would only be a slightly more polished reprise of the much-maligned RARE I and thus nicknamed it RARE II, which the Forest Service soon adopted as the official title.[35] Doug Scott, the Sierra Club's RARE II coordinator, supported the new study, calling it "historic" and "a great opportunity". He wrote Cutler that he would make "strong efforts" to help make RARE II a success. Scott continued to speak favorably about RARE II until the publication of the "draft alternatives" in June 1978. Disillusioned by what he considered to be RARE II's bias in favor of industry, he began calling it just another "quick and dirty" Forest Service attempt to dispose of the wilderness issue.[36]

The timber industry was also caught off guard by the announcement of RARE II. Once the study had begun, however, industry's attitude slowly began to change. Industry supporters participated in the public hearings and letter-writing campaigns.[37] Industry spokesmen expressed qualified endorsement of the RARE II results when Cutler announced them in January 1979. Their public statements may have been stronger if they had not feared that embracing RARE II would confirm the environmentalists' charge that it had been biased in favor of industry. Certainly, industry would have been more enthusiastic over RARE II if it could have been confident that its recommendations concerning 62 million acres of roadless land would have been quickly and thoroughly implemented by Congress.

The environmentalists' disenchantment with RARE II grew with the project's increasing ambitiousness. They were also distressed by its rapid timetable because they feared that Cutler and his assistants would be overwhelmed by data which would force them to rely almost completely on the recommendations of local Forest Service officials. At the beginning of the study, Cutler predicted that only the least controversial lands would be recommended for wilderness or nonwilderness status and that the bulk of the acreage would be left in the residual "further planning" category.[38]

As late as August 1978, only 4 months before the publication of the RARE II results, Cutler was saying publicly that one-half of the acreage would be placed in the further planning category. According to Cutler, it was during the next few weeks, as reports started coming in from the field offices, that "things started looking up and we realized we could make recommendations on more of the areas."[39] In the end the Forest Service "bit the bullet" and recommended 15 million acres for wilderness (5 million of which were on the Tongass National Forest in Alaska in a decision directly influenced by Cutler),[40] 36 million for non-wilderness, and the remaining 11 million acres for further planning. Most of the last category consisted of lands in California and the Chugach National Forest in Alaska, lands that previously had been committed to further study in the Montana Wilderness Study Act of 1977, or lands in the so-called "Overthrust Belt" of Wyoming where geologists suspected the existence of large reserves of oil and gas. During the interagency review of the draft recommendations, the Department of Energy strongly advocated releasing all potential oil and gas land to the non-wilderness category. Allocating most of the Overthrust Belt to further planning was a necessary compromise given the relative lack of data on the nature and extent of its oil and gas reserves.[41] The environmentalists maintained that the concentration of "further planning" lands in these categories gave a distorted picture of how much land RARE II had really set aside for further study. They believed they had lost the "allocation battle" because one-third of the land recommended for wilderness was in Alaska, where they had counted on doing well as part of the struggle over the Alaska Lands bill. Compared to the 36 million acres recommended for non-wilderness, the 25 million acres recommended for wilderness and further planning did not seem large when these factors were considered.[42]

After announcing the RARE II results on January 4, 1979, the administration declared that all nonwilderness lands would be "released" for other uses under the first cycle of forest plans mandated by the National Forest Management Act. Environmentalists had convinced officials in the Carter White House not to introduce an omnibus RARE II bill by arguing that RARE II had done enough damage and that they needed a free hand with Congress if they were to repair some of the damage.[43]

The timber industry, of course, saw things quite differently. The passage of individual wilderness bills and even the controversial statewide Alaska Lands Act, which was covered extensively by the national media, had demonstrated to industry that congressional representatives were often willing to vote for wilderness areas if these areas were not in their own States or districts. Wilderness had enough public support so that representatives could vote for wilderness bills in relative safety, knowing they would receive little criticism as long as the bill did not affect their constituents. In the eyes of the timber industry, wilderness had become a "cheap" environmental vote. It had to be made dearer by spreading the risks to more representatives. This meant promoting the idea of national legislation that would simultaneously decide the fate of all roadless areas in national forests.

In August 1979 former Congressman LLoyd Meads (D-WA), one of the architects of this strategy, explained its rationale in a speech to members of the Western States Legislative Forestry Task Force:

Now comprehensive legislation ties wilderness in 38 states together so that the guy from North Carolina... who traded his vote on Alaska wilderness is going to have to be voting on wilderness in his own district at the same time he is voting to lock-up forty percent of the forest in Alaska if you put it all in one package, and he is going to think about that,...[44]

During the previous 15 years the industry had seen what it considered to be a complete erosion of reasonable standards for wilderness designation. Organizations such as the Oregon Wilderness Coalition were championing the idea that wilderness should not be surrounded by a special aura but should be seen primarily as a way of preventing mismanagement by the Forest Service and the timber industry.[45] Although the Forest Service had recommended that 36 million acres in the RARE II inventory be released from further wilderness consideration, the industry feared that much of this land would eventually end up in the Wilderness System if the wilderness issue continued to be resolved on a piecemeal basis. In early 1981 the Forestry Affairs Industry Newsletter gave its view of the history of the wilderness movement.

Since passage of the '64 Act, wilderness criteria have been stretched wide enough to include any cutover, burned over plot without a ferris wheel or parking lot. If John Muir could see the land in East Texas recently demanded for wilderness he'd die again...

...Advocates demand wilderness not for scenic attributes, unique ecological offerings, or other intrinsic merits listed in the '64 Act. They want it simply because it's there. As a corporation won't limit profits thinking its shareholders won't settle for less, so preservation groups don't bridle their ambitions. The limit is what you can get. Preservation has become an end in itself, largely divorced from national needs...[46]

By 1979 the timber industry and other commodity interests appeared to be in a good position to push national legislation. The Iranian Revolution and subsequent cutoff of Iranian oil to the United States demonstrated again America's vulnerability to foreign energy supplies. For the first time, national opinion polls showed that Americans ranked energy above environmental quality in their scale of values.[47] The Carter Administration had created a new cabinet-level department, the Department of Energy, in order to deal with the so-called energy crisis, and, as previously mentioned, it had prevailed on the Forest Service to put potential oil and gas land in the Overthrust Belt into the "further planning" category.[48] Double-digit inflation and the rapid acceleration in the cost of housing prompted the Carter Administration to direct the Forest Service to deviate from its policy of not harvesting more timber annually than could be grown in a year (the sustained-yield even flow policy) so that the cost of building materials could be brought down. Industry could now hope to convince people that wilderness designation impeded efforts at attaining low-cost housing and energy independence.

Western Democratic politicians, such as Senators Frank Church (D-ID) and Warren Magnuson (D-WA), had been good friends of the wilderness movement. But their association with the Carter Administration's foreign and domestic policies and a renewed western rebelliousness over the way the federal government managed public land placed them in political jeopardy. They were no longer in a position to strongly support wilderness preservation.

In 1979 Congressman Tom Foley (D-WA), chairman of the House Agriculture Committee, who was also a western Democrat under strong attack from conservative opponents, introduced a bill that gave Congress time limits in which to designate RARE II wilderness areas. After the expiration of these periods, the lands would be "released" and would revert to management under regular national forest procedures. In 1980 his bill was altered to allow for the immediate designation of the 12 million acres that the Forest Service had recommended for wilderness. Neither of these bills stated that the Forest Service could never again recommend land for wilderness designation and thus did not threaten environmentalists as much as later bills would but they opposed them because they were afraid that the public and the Congress would assume that the wilderness issue had been totally settled. Perhaps just as importantly, they feared that their grassroots organizations would be destroyed or at least seriously demoralized if RARE II wilderness areas were designated without their participation.[49]

Congressman Foley negotiated with the environmentalists. He agreed not to oppose the Sierra Club's method of "releasing" roadless land. Because of his position as chairman of the House Agriculture Committee, which could assert jurisdiction over wilderness bills, his agreement on this issue was important to the environmentalists.[50]

In 1979 the State of California, over objections from the national environmental organizations, sued the federal government to prevent the development of 48 roadless areas that had been recommended for nonwilderness classification in RARE II. Judge Lawrence Karlton enjoined the development of these lands on grounds that RARE II had not provided site-specific impact data on the 48 areas as was required by the National Environmental Policy Act.

At the same time as Judge Karlton was reaching his decision, Democratic Congressman "Bizz" Johnson from northern California, an ally of the timber industry, and Phil Burton (Democrat) from San Francisco, the environmentalists' most powerful friend in Congress, were sponsoring separate California wilderness bills. The success of the State of California's lawsuit, however, meant that decisions to develop California roadless land could be legally challenged (as well as land in other States if local environmentalists used the California precedent) unless these decisions were removed from the jurisdiction of the courts. Thus began the search for "sufficiency" language, the objective of which was to insulate wilderness legislation from court challenges. Sufficiency language was soon linked to industry's desire for release language which would fix a time period after which the Forest Service could manage land studied but not designated as wilderness under regular national forest procedures. At first the environmental organizations had wanted to avoid any kind of release language and had urged the State of California not to sue because they feared a legislative "backlash" which might result in the Forest Service being forever forbidden to study released land for wilderness designation (permanent release) or forbidden to study it for several decades (long-term release).[51] If the Forest Service could never again study released land, wilderness activists would effectively be prevented from working with national forest supervisors.

Congressmen Burton, Johnson, and John Seiberling (D-OH), chairman of the House Public Lands Subcommittee, attempted to negotiate these issues but quickly decided to convene a larger meeting to see if they could reach a compromise among representatives of industry, environmental groups, and the Forest Service. After two meetings this group agreed on statutory language which would ensure sufficiency protection (protection against RARE II lawsuits) and which would release land examined but not designated as wilderness for one cycle (10 to 15 years) of the planning process required by the National Forest Management Act of 1976.[52] This was eventually called "soft" release. Congressman Seiberling believed this agreement would be kept by the timber industry because its representative, Dick Barnes, a counsel for an industry association set up to deal with RARE II, had participated in the negotiations. Barnes maintains that he had said he would take this agreement back to the industry for their approval "just as a labor negotiator takes an agreement back to the union membership." At a meeting in July 1980 of the industry's Public Land Withdrawal Committee Barnes told the industry leaders that he had gotten as good a compromise as could have been expected under the circumstances. Nevertheless, the committee disapproved of the agreement because the release period was not long enough. Barnes returned to Washington, DC and attempted to reopen negotiations but was rebuffed.[53] Senator Hayakawa's (R-CA) opposition killed the California bill in the Senate, but its release language was added to the Colorado and New Mexico Wilderness Acts, as well as the Alaska Lands Act, all of which passed at the end of 1980. ("Soft" release, "California" release, and "Colorado" release became interchangeable terms.) The fact that no senators or representatives had objected to the "soft" release formula as these bills moved through Congress convinced Congressman Seiberling that the timber industry had accepted it, despite Barnes' attempted re-negotiation.[54]

Whether the industry had agreed to the formula or had only temporarily avoided opposing it while an environmentally oriented administration was in power, they decided they could get a better deal on release after the 1980 election. Andy Wiessner, Seiberling's counsel on the Public Lands Subcommittee, believes the industry had broken its agreement but concedes it was "their right under our political system to do so."[55]

In February 1981 Senator Hayakawa introduced a bill co-sponsored by Senator James McClure (R-ID), the new Chairman of the Energy Committee and Senator Jesse Helms (R-NC), new chairman of the Agriculture Committee. According to the *Twin Falls* [Idaho] *Times News* "a dissident faction within the National Forest Products Association that had not been happy with Foley's or Burton's approaches was able to get its way on release language."[56]

Buoyed by the victory of Ronald Reagan, the timber industry decided to push more aggressively for permanent release. From its standpoint, the Hayakawa bill was a great improvement over the 1980 Foley bill since it did not create any wilderness but merely imposed deadlines on Congress after which all Forest Service roadless lands would be forever off limits to wilderness recommendations by the agency. If enacted, some argued it would have amended the 1976 National Forest Management Act, which required the Forest Service to reevaluate its lands every 10 to 15 years for all its multiple uses, including wilderness. During the next 3 years wilderness supporters in Congress, such as Congressman Seiberling, argued against amending the act on grounds that it should not be altered until it had more time to prove its usefulness.[57] The timber industry countered that Congress was always amending laws and that the National Forest Management Act was not special.[58] Of course, Congress would always have the right to create more wilderness, but both the timber industry and environmentalists knew it would be reluctant to do so without some kind of recommendation from the Forest Service. Steve Forrester, a reporter and close follower of the wilderness movement in the Pacific Northwest, described the optimistic mood of the timber industry during the early days of the Reagan administration:

If you had asked the timber lobbyists who crowded into the Senate Energy Committee hearing room in February 1981 to make predictions about what legislation would be enacted during the next four years of the Reagan administration, it's a sure bet they would not have predicted enactment of Oregon and Washington state wilderness bills. With Senator James McClure, R-Idaho, running the Energy Committee, Ronald Reagan in the White House and John Crowell at the Forest Service, the timber lobby was in heaven and wilderness was the farthest thing from their minds.[59]

The Senate Public Lands Subcommittee of the Energy Committee scheduled its first hearings on the Hayakawa bill during a Senate recess. The environmentalists feared this was the beginning of a legislative "freight train" that would sweep through the Republican Senate and then overcome opposition in the Democratic House, as was happening with the administration's tax and budget bills. Senator Dale Bumpers (D-AR), former chairman of the Public Lands Subcommittee, was out of Washington during the hearings and regretted not having been able to attend. He requested and was granted a new hearing, which Tim Mahoney of the Sierra Club says was the club's first "foot in the door" in its fight against the Hayakawa bill.[60] At this hearing Rupert Cutler, now National Audubon vice president, predicted that the Hayakawa bill would simply "gum up the legislative machinery" and that RARE II wilderness legislation would be completed by the end of 1984.[61]

The environmentalists worked with Senator Walter Huddleston (D-KY) of the Agriculture Committee who secured an agreement from Senator Helms that there would be adequate notice before the committee held its hearings on the bill.[62] (The Agriculture Committee never held these hearings.) These delaying actions prevented the bill from quickly passing out of committee and going to the Senate floor where the environmentalists feared it had a good chance of being passed.

The environmentalists wanted very much to obtain at least the neutrality of Senator John Melcher of Montana, the only Democrat who was a member of both the Agriculture and Energy Committees. When he was in the House, Melcher had been a leader in drafting the National Forest Management Act. It was hoped his interest in that legislation would prevent him from supporting permanent release. If he were to ally himself with the sponsors of the Hayakawa bill (which he never did), the environmentalists feared his influence in both committees might tip the balance against them. At the same time, the environmentalists were attempting to gather support from several senators, especially Republicans, so that they could "throw sand in the gears" of the bill. Thus, even if it passed the Senate, enough doubt would have been created about the value of the bill to increase the possibility of its defeat in the House. Four years later a staff member of the Senate Energy Committee revealed to Tim Mahoney that enough questions had been raised at the hearings to doom chances of the bill getting out of committee. This was not clear to the environmentalists at that time and Mahoney speculates hindsight may have helped the staff member reach this conclusion. In any case, the environmentalists lobbied against the bill as if it posed a serious threat. They were aware that the bill had not been adequately drafted and that it contained many apparent contradictions, but they were reluctant to stress these problems for fear of becoming trapped in an effort to improve its technical coherence. Their goal was to defeat completely the idea of permanent or long-term release and not to make it more compatible with other forestry legislation.[63]

Within a few months of the Senate hearing it became clear that the Hayakawa bill would not bask in the legislative glow of the "Reagan Honeymoon", despite support from the administration, important senators, and industry. The Eugene *Register Guard* in the important timber-producing region of western Oregon described the stalemate that existed at the end of 1981. An editorial pointed out a problem supporters of the Hayakawa bill found difficult to overcome - i.e. even if the timber industry had not reneged on the 1980 "soft" release, many people believed it had.

The timber industry took a risk earlier this year when it backed out of a compromise made at the end of the 96th Congress and moved to support a new bill sponsored by S.I. Hayakawa. . . Within one month after Hayakawa's bill was given hearings last April by the Senate Energy Committee, it was clear to most observers that Chairman James McClure, R-Idaho, didn't have the votes to move the legislation. The last seven months have been fraught with rumors that McClure was trying to achieve compromise and consensus to make the bill palatable. But nothing seems to have occurred, and McClure's committee is no closer to resolving the release language issue, hence the wilderness issue...

Still, there is no evidence that industry is willing to compromise. 'There is no softening in the industry,' says Gene Berghoffen of the National Forest Products Association. 'We are no more willing to compromise now than we were when the bill was introduced.' Industry has acted as though release language were almost a theological question on which there can be no deviance, no compromise. 'It must be obvious to anyone by now that the Hayakawa bill isn't going anywhere,' says an aide to [James] Weaver [Democratic Congressman from Eugene], 'but industry is trapped by its own rhetoric.'[64]

At the end of 1981 Senators McClure and Melcher proposed a compromise that would have been similar to the second Foley bill of 1980, although with slightly fewer acres to be immediately designated as wilderness. The Sierra Club was encouraged that permanent release had been eliminated from the proposal, but both the club and The Wilderness Society continued to oppose any kind of national wilderness bill. They also objected to some features of the proposal that they felt would amend the National Forest Management Act and establish unreasonable deadlines on wilderness study areas. A Wilderness Society staff member observed that "McClure and Melcher have tried. They can go back to the timber industry now and say we have tried our hardest. Eventually, we hope the whole release idea will just blow away."[65]

The release idea did not blow away, but in the Wyoming bill introduced in early 1982 it mutated into what was a slightly less threatening form for the environmentalists - so-called "time certain release" that would have prevented the Forest Service from recommending any released land for wilderness in a land management plan made before the year 2000. Because most "second-generation" land management plans would be revised in the late 1990's, year 2000 release meant that effectively land would remain released until the first plans of the 21st century were written or sometime between 2010 and 2015. Many environmentalists saw this as tantamount to very long term release.[66]

These facts contained a potential hazard for the environmentalists. Release was an issue not readily understandable by the general public or even by many wilderness activists. In 1984 Steve Forrester reported with only slight exaggeration that just three people really understood release - Peter Coppelman of The Wilderness Society, Andy Wiessner of Congressman Seiberling's staff, and Bill Brizee of the Department of Agriculture.[67] Many Sierra Club members did not follow the intricacies of the release debate but merely accepted on blind faith what their leadership told them. But this was not true of the Oregon Wilderness Coalition and some other environmentalists, who felt they had little stake in the release problem.

The Coalition was becoming increasingly estranged from the Sierra Club and The Wilderness Society, which placed resolution of the release question ahead of the passage of any single State wilderness bill.[68] Many legislators also did not understand what to the environmentalists were fine points of principle and would have been willing to settle for a compromise somewhere between permanent national release with no wilderness designation and "soft" release in statewide bills. The environmentalists had important legislative allies, who thoroughly understood their position on release. But even this support might have been inadequate in the face of a public becoming impatient over what seemed to be pettifogging differences and the timber industry's argument that the environmentalists' intransigence on release was threatening timber supplies and jobs.

The release question and wilderness might have become boring issues unable to hold public attention if James G. Watt had not become the secretary of the interior. Watt did not have any direct responsibility for managing national forest wilderness but through his department's authority to grant leases and patents for mining on all public lands he influenced the outcome of the wilderness debate. Because he enjoyed ideological debate and strongly believed in the development of natural resources, he revived the wilderness debate and gave the wilderness organizations issues with which to arouse the public.

The 1964 Wilderness Act contained one provision the environmentalists had accepted only as the last and hardest compromise needed to pass the act - an extension of the public mining laws in wilderness areas until December 31, 1983. But although mining companies technically had the right to explore and lease minerals in wilderness areas, they had not done so because of strict Forest Service regulations, negative public opinion, and the costs of exploring in remote areas where prospects were unknown. Consequently, for several years the federal government had imposed a moratorium on the processing of claims. According to Howard Banta, the Forest Service's chief minerals expert:

Industry saw the handwriting on the wall. They were simply not going to be allowed - even though the law permitted it - from the standpoint of public opinion to really explore. And there was very little certainty of being able to develop anything they would find. As a consequence... there was very little exploration done in statutory wilderness...[69]

As the mining deadline approached, the mining companies began to realize they would forever be excluded from wilderness areas unless they managed to establish claims before the end of 1983 or got Congress to extend the deadline. Bills to extend the mining deadline for 10 or 20 years were introduced but never stood a good chance of passage. Thus, the companies had to rely on establishing claims by convincing the government and the public that they had a legitimate right to enter wilderness areas. They mounted a campaign to get the federal government to lift its moratorium.

James Watt, before he became secretary of the interior, headed the Mountain States Legal Foundation, which had been patterned after the so-called public interest environmental law firms. Although Mountain States was organized like these other groups, its founder, Joseph Coors, the conservative owner of the Coors Brewery in Golden, CO, had just the opposite purpose in mind. His foundation was set up to combat what he believed to be excessive governmental regulations hampering industrial growth.[70] In 1980 Mountain States sued the federal government on grounds the Forest Service was illegally holding up two mineral leases. The district judge ordered the secretary of the interior to report the withdrawal of the lands to Congress within 20 days or cease withholding them and to promulgate rules relating to the bases on which oil and gas applications could be rejected, approved, or suspended. Mountain States saw the decision as a vindication of its position. The environmentalists, however, chose to interpret it only as an order to the government to decide either for or against the claims.[71] Watt's role in the lawsuit made it almost inevitable that he would become directly involved in the wilderness leasing issue when he became secretary of the interior.

The election of Ronald Reagan brought the morale of the Sierra Club's leadership, which had been slipping ever since the energy crisis of 1979, to its lowest ebb. When James Watt was nominated for secretary of the interior, the top leadership of the club was reluctant to oppose him although they suspected he would be their strongest opponent. A small-scale mutiny within the club's Washington office resulted in a reversal of the tentative decision to acquiesce and the club began actively to oppose Watt's nomination. Even though that attempt received little overt political support, it put the club into a more assertive mood and brought it out of the defeatist mentality produced by the initial conclusion that President Reagan had won decisively and should be allowed to appoint whomever he wanted. A similar fight to block the appointment of John Crowell as assistant agriculture secretary for natural resources, overseeing the Forest Service, improved the environmentalists' morale still further. The environmentalists argued that Crowell's former position as chief counsel for Louisiana Pacific, the largest buyer of federal timber, created a possible conflict of interest.

Crowell, who was from Portland, OR, had the support of his senator, Republican Mark Hatfield. Crowell assured Congress that he would not have a conflict of interest. Nevertheless, the environmentalists had created enough doubt to convince 25 Senators to vote against his confirmation.[72]

Soon after becoming secretary of the interior, Watt began some activist policies, such as leasing oil sites off the California coast, that provoked strong controversy. When the issue of mineral leasing in wilderness came up, however, Watt claimed with some justification that he was merely doing what the law required of him. His environmental opponents, on the other hand, believed that Watt could have followed the example of previous secretaries and found legal and administrative ways to stop leasing if he had wanted to. The Sierra Club had obtained a high-level Interior Department memorandum stating that one of the Department's goals was "to open wilderness areas." This document was given to the press as evidence of Watt's active support for mineral leasing in wilderness. Secretary Watt denied that he had personally directed the department to open wilderness areas.

The first test for mineral leasing in wilderness areas came in the Bob Marshall Wilderness in western Montana, located in the geological formation known as the Overthrust Belt, which geologists believed contained substantial supplies of oil and gas. This 1.5-million-acre complex (including two contiguous wildernesses) was named after one of the founders of the modern wilderness movement and was often called the "flagship of the Forest Service wilderness fleet."

A Denver-based energy company had applied for a permit to conduct seismic tests using explosive charges in "the Bob", as it was called for short. Because of its exalted stature, the Bob permit request at once assumed symbolic importance and aroused intense emotion among the many ardent wilderness supporters in western Montana who organized the Bob Marshall Wilderness Alliance to stop "the bombing of the Bob." If a company could get a permit to work in the Bob, one of the most beloved of national forest wilderness areas, then it could probably get permits in many other areas. The *Missoulian* newspaper in Missoula, Mt, portentously editorialized:

> *Many defenders of the Bob all along have suspected that the Denver Firm's controversial proposal is less motivated by a desire to find oil or gas than it is to test whether energy exploration in wilderness will be tolerated. By picking the Bob, they picked the best. If only they can explore there, they can explore anywhere.*[73]

Tom Coston, the regional forester in Missoula, denied the permit request because the company had committed a technical error by not applying under the proper mineral leasing laws. The agency's Washington office directed the regional forester to reach a decision based on the appropriate laws. Coston again denied the permit on grounds that seismic testing would disturb delicate grizzly bear habitat and otherwise damage the ecological integrity of the Bob.[74] But these administrative decisions were not enough to protect the area because they were subject to further appeal.

At first, the national wilderness organizations were unsure of how to deal with this situation and reacted passively to the events unfolding in Montana. Congressman Pat Williams (D-MT) took the initiative and began to work with Andy Wiessner of the Public Lands Subcommittee to find some way to withdraw the Bob Marshall Wilderness from the possibility of seismic testing and mineral leasing. They believed they had found at least a stopgap measure in an obscure section of the Federal Land Policy Management Act (FLPMA) of 1976, which was the "organic act" for the Bureau of Land Management in the Department of the Interior. This provision permitted the Interior Committee to withdraw land from commercial activities if a public emergency existed.[75] It had only been used once before to protect the town of Venture, CA, from the possibility of well contamination from uranium mining on nearby federal land. As the newspapers later reported, this was certainly a novel interpretation and stretched the concept of the separation of powers to the limit. Republicans on the Interior Committee asserted that it would be unconstitutional to use it in this case because all of the administrative remedies had not yet been exhausted. Interior Chairman Morris Udall (D-AZ) admitted it was a novel use of FLPMA but charged it was the price Secretary Watt had to pay for his "confrontational administration." Interior Committee lawyers were also dubious about its legality but the Democrats on the Committee, faced with no alternative and strong public pressure "to save the Bob" decided to use it anyway.[76] Tim Mahoney has described this first round in the congressional attempt to prevent oil and gas leasing in wilderness.

> *We wanted the Committee to pass a resolution withdrawing the Bob from mineral leasing but not mining because the hard-rock miners are very tough - to try to cut the Mining Congress away, which we did with some success, and just work on the petroleum guys, who, although very rich, had a lot of irons in the fire. We had a real dog-fight in the Interior Committee. It was vicious and long and there was a united Republican opposition to it. We had to work real hard to line up all the Democrats. Udall worked hard to line up some of the softer Democrats [the final vote was 23 to 18]. We got it through and it was immediately challenged by Mountain States Legal Foundation, Watt's old outfit. . . We had temporarily saved 'the Bob' but not much else.*[77]

Secretary Watt personally delivered a letter to Chairman Udall stating he would comply with the committee's resolution to close the Bob Marshall, despite serious misgivings about the constitutionality of the action. Then came charges in the press that Watt had retaliated against Udall by threatening to stop construction of a water project in his district and through the back door had made it known that he would cooperate with Mountain States and Pacific Legal Foundation, which had joined the suit with Mountain States. The Justice Department, which had announced it would not defend the committee against the suit, released a letter from Pacific Legal Foundation stating that they "were informed that the filing of this lawsuit was received with favor by the involved officials of the executive branch." No evidence was actually produced that showed Watt had conspired with Pacific Legal Foundation but the incident served to heighten controversy surrounding his administration.[78] The Billings (MT) *Gazette* noted the convoluted turn the case had taken:

> *The court case has become a curiosity. Watt is the defendant, defending an action he opposed. He closed the Bob Marshall on the order of the House Interior Committee after mak-*

ing it clear that he wanted to increase mineral exploration in wilderness areas. Attacking Watt in the suit are the Pacific States Legal Foundation and Watt's old Mountain States Legal Foundation. Both are generally supporters of Watt's policies and consider Watt a champion of their own beliefs. Siding with Watt in the case are his avowed enemies, including the Sierra Club and the Bob Marshall Alliance. They contend Watt was right in closing the wilderness, despite his convictions.[79]

With the Bob Marshall temporarily protected, the action shifted to New Mexico where, for the first time, local Interior Department officials had actually approved a mineral lease in the Capitan Wilderness, which had been created by the New Mexico Wilderness Act of 1980.

When this news reached the Sierra Club's Washington Office, the staff members at first were perplexed over how to deal with a threat to a relatively new and obscure wilderness area. But then they discovered they had been handed a potential public relations windfall. The Capitan Wilderness was not just "any wilderness. It was the wilderness where Smokey Bear had been found and Smokey symbolized not just the Forest Service but forests in general."[80]

Manuel Lujan (R-NM), the ranking Republican on the Interior Committee (and later secretary of the interior in the Bush Administration), did not have the Capitan Wilderness in his district but did feel a responsibility for all the wildernesses in his State. He, like his fellow Republicans, had voted against the Bob Marshall resolution, but he now was angry that a lease had been granted by local officials without his knowledge. His first reaction was to propose a resolution banning all leasing in wilderness areas. After meeting with Watt, however, he dropped that idea and settled for an assurance that no leases would be granted by local Interior Department officials and that environmental impact statements would be prepared in all cases. This agreement was totally inadequate, as far as the environmentalists were concerned, and prompted Peter Coppelman of The Wilderness Society to observe that Watt's agreement to follow existing regulations merely meant he had promised to "stop beating his wife."[81]

Chairman Udall would not drop the matter and sponsored a resolution calling on Secretary Watt to refrain from issuing any mineral leases until June 1982 so that Congress would have time to solve the problem. His resolution passed overwhelmingly (41 to 1), whereas a resolution by Phil Burton of California to ban immediately all leasing in his State lost by one vote. (A failure to coordinate closely with Udall was probably the reason Burton's resolution did not pass).[82] Secretary Watt then promised to delay issuing any leases until the end of the 1982 session of Congress. Watt continued to maintain that his hands were tied and that he would have to issue leases unless Congress passed legislation directing him otherwise. In a memorandum to Doug Scott in the Sierra Club's San Francisco office, Tim Mahoney sounded the optimistic note which was now appearing among the environmentalists.

Among the positive benefits of all of this was some of the best pro-wilderness rhetoric ever to be heard in the Interior Committee... The level of interest was simply fantastic; the hearing room was jammed with the press and spectators... It clearly puts a wet blanket on the idea of extending the 1984 deadline...

Although we are clearly disappointed that we did not get the full withdrawal of all areas in the wilderness system, the other side must view all of this as loss as well. Overall we are far ahead of where we were two weeks ago.[83]

The leasing battle was also stirring intense interest in Wyoming, Watts' home State, where there were requests to lease in the half-million-acre Washakie Wilderness. Over the years citizens of Wyoming had not shown as much support for wilderness as citizens in the neighboring states of Colorado and Montana. As the most rural of the Western States, Wyoming did not have many prowilderness urbanites but it did have a large percentage of its population, mainly ranchers, who derived their livelihoods from the national forests. This often meant there was considerable resistance to the creation of new wildernesses, which some people feared would threaten existing economic relationships. But the leasing dispute showed that while Wyomingites might be ambivalent about designating more wilderness, they were eager to protect what they already had. To many people in the State, wilderness areas represented islands of stability in the midst of the rapid change being brought about by the development of the State's coal and oil deposits. The Wyoming congressional delegation, realizing the depth of public opinion, came out against leasing. The *Denver Post* noted one of the ironies created by the leasing issue:

In the environmental wars between diggers and conser-

vationists, Rep. Richard B. Cheney Jr., R-WY is usually found on the side of the diggers. He fervently supports oil and coal development in his state. He is an admirer of Interior Secretary James G. Watt's dig-it-up now approach to energy resources. He is the despair of environmentalists, who calculate he voted their way only 12 percent of the time - compared with a House average of 48 percent - during his first term. Yet today Cheney and other conservative Western Republicans are unlikely rebels in the touchiest environmental fight of the year, the clash over mineral leasing in the nation's designated wilderness areas.[84]

The expression of western public opinion was beginning to influence many legislators. It was becoming clear that it would be politically impossible to allow leasing simply on the grounds that the Wilderness Act did not prohibit it. Early in the next session of Congress Secretary Watt appeared on the "Meet the Press" television interview show to announce that he had decided to offer a bill that would immediately ban leasing in wilderness unless the President declared the resources were needed for a national emergency. He went on to say that the ban would stay in effect until the year 2000 and then "let's have Congress revisit that issue in the next century." Copies of the bill had not yet been made available and the questioning of Watt did not draw out many details of his proposal. Consequently, there was at first some confusion within the ranks of environmentalists. The Wilderness Society's executive director called it a "turnaround" while the Sierra Club's spokesmen had heard Watt use the word "certainty", which they believed was a "code word" for permanent release. They told the press they suspected that Watt's proposal contained a "Trojan Horse".[85]

The bill was made public 2 days later. Environmentalists were disconcerted to learn that the bill would reopen wilderness areas to leasing in the year 2000. It was this aspect that the press generally reported on. But the bill also contained all of the provisions of the Hayakawa permanent release bill, as well as a section that would allow the President, without congressional concurrence, to release Bureau of Land Management wilderness study areas if they were not found suitable for wilderness designation. The trade of a temporary leasing ban for permanent release was one the environmentalists and their congressional allies were unwilling to consider. They immediately denounced the bill. Several reporters who felt they had been tricked by Watt's seemingly contradictory announcement on "Meet the Press" continued to inform their readers about the story.[86] A *Los Angeles Times* editorial summed up the feeling of most journalists familiar with the issue.

> The original law allowed applications for leases for 20 years; no Interior Secretary had ever encouraged leasing in the wilderness. Watt did. So his concern deals with a problem that he created...
>
> The Interior Department draft bill would set deadlines for Congress to make up its mind on the new [wilderness] acreage. If Congress missed the deadlines, the land would be up for grabs again.
>
> That is a sly notion. Congress is not good at meeting deadlines. .... Watt should acknowledge that he is wrong on the issue and stop trying to force his way into the wilderness disguised as a friend of nature.[87]

The public was generally opposed to the bill, which was reflected in the failure of any senator to sponsor it. Although Watt's bill died quickly, it showed that wilderness leasing eventually would be banned. Would it be tied to a national release provision, as the proponents of permanent release desired, or would the two issues be separated, as the environmentalists wanted?

In June 1982 Congressmen Seiberling and Lujan offered a bill that would permanently ban wilderness leasing. The bill passed the House after a section had been rewritten to allow claims to be filed for hardrock minerals in eastern wilderness areas until the deadline of December 31, 1983. (Eastern public lands had been acquired and had not been part of the public domain; therefore, hardrock mining in the East came under the mineral leasing acts and not the Mining Law of 1872.)[88]

In the Senate Henry Jackson (D-WA) had become involved with the issue because of requests to lease in the popular Alpine Lakes Wilderness near Seattle. Jackson was an extremely hard worker and a formidable gatherer of support for bills he sponsored. Within days of introducing a bill to ban leasing in all designated wildernesses, he had enlisted 53 co-sponsors, many of whom he had convinced while talking in the Senate locker room.[89]

As fall approached, it was not clear whether Secretary Watt's temporary ban on processing leasing applications would extend through the lameduck session of Congress after the November elections. The

Jackson bill was blocked in the Interior Committee by Senator McClure who objected to the protection it gave to wilderness study areas and the absence of a national release provision. Assistant agriculture secretary, John Crowell, expressed the administration's support of a leasing ban in exchange for national release.[90]

In order to circumvent the Senate Interior Committee, proponents of the leasing ban decided to attach a rider to the Interior Appropriations bill prohibiting the Department from spending any money to process claims for exploration or leasing. This rider was passed as part of a continuing budget resolution, which lasted through the lameduck session of Congress. In December the same rider was attached to the Interior Appropriations bill for Fiscal Year 1983. Senator McClure tried to add a national sufficiency provision in the Interior Appropriations Subcommittee, which he chaired, but was defeated. The appropriation ban was good until July 6, 1983, 6 months short of the December deadline. Secretary Watt, however, agreed not to process any leasing claims before the deadline because, according to an anonymous high ranking official; "The American people have reached a social decision about the wilderness [leasing] issue."[91] Thus, the environmentalists finally had succeeded in separating the release and leasing issues. Russ Shay, the Sierra Club's California and Nevada representative, told his chapter's members:

I remember back a few years ago when the upcoming 1984 deadline for new mineral leasing or claims in wilderness areas looked to me like a sure legislative armageddon in which the oil and mining companies would overwhelm wilderness supporters. We have turned the tables.[92]

According to Mark Reimers, the Director of the Forest Service's Legislative Affairs Staff, the agency viewed "release" and leasing as separate issues which did not really affect each other.[93] The national environmental organizations, however, saw both as related battles in their overall wilderness campaign. They felt they had emerged from their wilderness leasing victory more confident of their position on the release question and their ability to influence wilderness legislation. The timber industry, on the other hand, was in a severe economic recession which was hampering its lobbying efforts on release. By late 1982 a large segment of the industry was concerned less with release than with getting "relief" from federal timber contracts purchased when the cost of stumpage was high. These companies, mostly in the Pacific Northwest, were selling their products at prices that made it uneconomical to harvest the federal timber previously purchased. The effort to get legislation to escape from these contracts diverted resources away from the release battle. It also made it more difficult to assert that wilderness designation withheld needed timber supplies when many companies were trying to turn back supplies they already had. The industry argued that timber "relief" was a short-term problem caused by cyclical economic conditions whereas wilderness designation affected long-term supplies.[94] But this distinction was not always clear to the public and the environmentalists were able to exploit the industry's difficulties in their media and fundraising campaigns.

Both sides were trying to get a Western-State bill through both Houses of Congress, with the kind of release language they favored. All the groups agreed that the first successful bill (not counting the Colorado and New Mexico bills of 1980) would set a precedent for all subsequent bills because it would be impractical to ask the Forest Service to follow a different procedure in each State. In 1982 the Senate passed a Wyoming bill that had long-term release supported by industry. If the House also were to pass a Wyoming bill, industry hoped a House-Senate conference would result in a compromise containing at least some of the provisions they wanted. Chairman Seiberling, however, had enough political support to keep the bill "bottled up" in his subcommittee.[95]

The environmentalists were concentrating their strategy on the three largest timber-producing States in the West - California, Oregon, and Washington. In August 1981 Tim Mahoney outlined the cautious strategy which the Sierra Club, with some modifications, followed over the next 2 1/2 years.

In the past, we have called our approach the 'leapfrog' strategy. We moved California first because it could best move in the House and pave the way for Washington on release. Then, as California bogs down in the Senate, we would try to move the Washington bill through both House and Senate. If Washington goes through, the odds on Oregon increase. The success of all three are interrelated, and if any of them fail badly, it hurts the others and our overall strategy.

. . . Passage of the California bill is an excellent start [refers to the 1981 House passage], but we cannot stop there. We face a danger that if any of our state-by-state bills erupts in controversy over on the House side, the timber industry, as well as Senate leaders and even some nervous friends like Representative Foley,

will take it as a sign of our weakness and the weakness of our approach."[96]

In 1982, Congress passed wilderness bills for Florida (vetoed by President Reagan), Indiana, and West Virginia, which contained the 1980 Colorado release language. Senator McClure did not oppose these bills because he considered them to be in a different category from the big western bills and because they were handled by the Senate Agriculture Committee. To the environmentalists, however, it was important to "trickle" a few wilderness bills through Congress to show that "the system still worked."[97]

According to Scott Shotwell of the National Forest Products Association, the House Public Lands Subcommittee vote on the West Virginia bill was especially important.[98] Unsure of having enough votes, Seiberling postponed two scheduled mark-ups of the bill. A last-minute amendment allowing the Governor of West Virginia to suspend certain standards from the Clean Air Act enabled the bill narrowly to pass out of the subcommittee.[99] If the bill had failed, it would have given that sign of weakness, which Tim Mahoney had warned about in 1981.

By the fall of 1983 there were indications that the stalemate over release was beginning to break. The previous summer Senator Mark Hatfield had held field hearings on an Oregon wilderness bill after having stayed aloof from the wilderness issue for 3 years. The environmentalists had been counting on Senator Henry Jackson to push through a Washington bill that would break the release deadlock. His death in August 1983 made Hatfield's return to wilderness politics even more significant.

Also that summer the Forest Service testified for the first time on behalf of a bill (Utah) that contained some land not recommended in RARE II.[100] In October deputy assistant agriculture secretary, Douglas MacCleery, testified that the administration would be willing to accept "permanent or long-term release", a softening of the department's previous position that only permanent release would do.[101]

Forest Service officials also began to talk to various Senators and Representatives about certain technical deficiencies that the agency had found in the 1980 release language. Of primary importance was the meaning of the word "revision" in the "soft" release formula. If a forest plan were amended, would the roadless land have to be reconsidered all over again for wilderness designation or could that wait until a whole new plan was written? Senator Jesse Helms asked the Congressional Research Service (CRS) to study the question. Its first report stated that it was quite possible a court could construe an amendment as being a revision requiring reexamination of roadless land. Peter Coppelman, The Wilderness Society's counsel, disputed this conclusion, which he claimed was based on an inadequate reading of the Colorado Act's legislative history. A subsequent CRS study reached the opposite conclusion but some legislators had begun to doubt the adequacy of the "soft" release formula.[102]

Senator John Melcher was one of the first Senators the Forest Service spoke to about these problems. Melcher took a personal interest in the issue and brought it to the attention of Congressman Seiberling, who had been his colleague on the House Interior Committee in the 1970's. Melcher respected Seiberling's ability to draft legislation and was confident that he would quickly grasp the problem and see the need to modify the Colorado language.[103] According to Tom Thompson of the Forest Service's Legislative Affairs Staff, Melcher helped create a climate that led eventually to the compromise over release.[104] (The Wilderness Society later thanked him for keeping the discussion from straying beyond the bounds of the National Forest Management Act into permanent or long-term release.)[105]

The environmentalists, of course, were aware that these discussions were taking place and were afraid that if their congressional allies accepted any of the technical objections, the timber industry and its congressional supporters would have opened the door to other concessions. In a December 1983 letter the Sierra Club and The Wilderness Society expressed their fears to their most strategically placed ally, Congressman Seiberling:

While interest groups have a right to be concerned about the future, it is impossible to design any legislation that will cover every hypothetical possibility that might conceivably be encountered for decades. From a strategic standpoint, it is to the timber lobby's advantage to continue to seek new 'clarifications' even as old problems are explained away. If the problems are dismissed, the industry has lost nothing. If they raise new doubts about the old language, the lobby is aided in pushing a renegotiation. The renegotiation would only go in one direction, towards statutory-length forest plans.

We prefer a strategy where the other side has to put specific new language addressing specific problems and we get to pick at their language and its inadequacies rather than guessing in advance what 'clarifications' the Senate might accept...

The ultimate agenda for the timber interests is to see that wilderness is not reconsidered for a long time... They would like the release language to thwart any legal challenge on any grounds to the initial Crowell-plans and, more importantly, to ensure that the forest plan has a statutory life so that it cannot be revised by a subsequent Administration... We would be giving the timber industry an advance statutory commitment for a plan of Secretary Crowell's choosing - which we have never seen.[106]

On the same day as this letter was being written, the Oregon Natural Resources Council (previously called the Oregon Wilderness Coalition) had filed a lawsuit asking the court to stop the development of any RARE II lands in Oregon based on Judge Karlton's 1980 California decision. The Council had threatened for 3 years to sue but had been dissuaded by the national environmental organizations who feared a legislative backlash that might result in permanent release. The Council had finally lost patience when Senator Hatfield did not introduce a bill in 1983 and decided to go ahead with a suit knowing that the Sierra Club and The Wilderness Society would publicly disavow it.[107] If successful, the lawsuit would have had, at least temporarily, a severe impact on Oregon's timber industry.

According to Tom Imeson, Senator Hatfield's principal assistant on the wilderness issue, the lawsuit did "not alter the Senator's timetable" but it did give him an argument with which to persuade his colleagues that an Oregon bill needed to be passed quickly.[108] If the Council's action had come 2 years earlier when the timber industry was politically stronger, it may have resulted in stronger efforts to pass permanent release. But in late 1983, with the Hayakawa permanent release bill long dead and Wyoming long-term release moribund, the Council's suit helped nudge Congress towards a release compromise acceptable to the environmentalists.

Senator Hatfield had not met with the Sierra Club for several years but in February 1984 he granted a meeting to Tim Mahoney, Jim Blomquist, the club's northwest representative, and Ron Eber, the club's volunteer leader in Oregon, to discuss his upcoming bill. At that meeting he promised to support the 1980 language. According to Mahoney:

Eber, Blomquist, and I had prepared a long speech to give to Hatfield on how we had appreciated working with him and that release language was very important and how we cared about these several areas. Instead we were ushered into his personal office and there was nobody there, so we sat in the corner of this huge office. We were waiting for ten minutes. He's a master of the psychology of these sorts of things. And then he sort of breezed in and before we could say a word, he said: 'It's California release and I can't tell you everything about the bill but it won't be a million acres, a million acres is too much but it will be a little less than a million acres with room for some negotiations between the time I introduce my own bill and the time we actually mark it up and get it out and have the final bill.' Then we asked him if he had consulted with the House and he said 'not really.' We tried to warn him that he might have trouble with the Senate Energy Committee, that they had never passed a bill with soft release for one planning cycle and he said 'well, they passed my bill in 1979 and they'll pass my bill now. I'll support Senator McClure and Senator Wallop on their bills and they better support mine.'[109]

By March wilderness bills for New Hampshire, Vermont, Wisconsin, and North Carolina had been prepared and were ready for action by the Senate Agriculture Committee. All contained the 1980 soft release language. The North Carolina bill had been worked out by Congressman Jamie Clarke (D-NC), Forest Supervisor George Olson, the North Carolina Department of Natural Resources, and the local Sierra Club and timber industry. The bill easily passed the House after receiving the support of all of the State's representatives. Senator Helms, however, had not yet agreed to support it. The other three State bills had the unanimous support of their delegations.[110]

It had been reported that Senators Helms and Melcher would try to alter these bills so that forest plans would run for a fixed 10- or 15-year period during which roadless land could not be reexamined under any circumstances. The Sierra Club and The Wilderness Society criticized this plan as being too "inflexible", although they knew it was much closer to the 1980 formula than the Wyoming bill, which would have put off wilderness reconsideration until the second decade of the next century. At the markup session on March 28, Senators Helms and Melcher, as well as Forest Service Chief Max Peterson, said that release was a national issue and that problems in the 1980 formula needed to be corrected.

Senator Patrick Leahy (D) of Vermont and Agriculture Subcommittee Chairman Roger Jepsen (R) of Iowa argued that the local delegations should be allowed to have language they had agreed on. Senator Orin Hatch (R-UT) said he wanted "hard" release for the Utah bill but favored allowing each delegation to have the bill it desired. Close followers of the release question, however, believed it was very unlikely that Congress would pass different release formulas. They suspected that the States rights arguments would work only for the first bill to pass Congress.[111]

Leahy prevailed and the bills were passed out of the committee on a 10 to 0 vote after Senator Helms had requested that the North Carolina bill be allowed to accompany the other three. According to Bill Brooks of the Agriculture Committee staff, Senator Helms did not view this as a serious matter because he put holds on the bills until the Senate Energy Committee had reached an agreement on release.[112] The Sierra Club's Tim Mahoney, however, saw it as an important event and a signal to Senator McClure that he did not have enough support to pass long-term release.

> *Nevertheless, the unwillingness of the Agriculture Committee to override the senators from local states was a victory. The unwillingness of Helms and Melcher to put it to a vote was a victory and the unwillingness of Helms to go it alone in North Carolina and possibly lose his bill was a victory. He blinked. All in all this was a very good day for all of us.*[113]

At the end of March the *Congressional Quarterly* reported that Senator McClure had agreed to accept soft release in the Washington bill and that Senator Hatfield was "pushing just as hard for his bill." The *Quarterly* predicted that the "moment of truth" would come April 4.[114] That moment actually came April 11 in "a dramatic showdown" between Senators Hatfield and Dan Evans (R-WA) and Senator McClure. Senator Hatfield appealed to the principal of comity and alluded to his powerful position as chairman of the Appropriations Committee.

> *Now, Mr. Chairman, I think unlike any other of the western states at this moment, we are in a very unique situation in Oregon in that we have had suits filed that are now being litigated. . . And considering the basic economics of Oregon is related to the timber industry, we are in a very urgent situation. . .*

> *I have a very strong feeling that, in relation to the release language, that we're now playing a game of hostage; that one state that may have a resolution of that problem cannot move because some other state hasn't had a resolution of their problem.*

> *Now, as Chairman of the Appropriations Committee, I have tried to accommodate the members of the Senate in literally hundreds of amendments. . . And I have tried to perform that same help and assistance with senators on this committee.*

> *Now, Mr. Chairman, I am a little bit impatient today because I've been ready to move on this bill for quite some time. . .*[115]

Senator Evans expressed his state of extreme readiness by referring to a logjam metaphor used earlier in the colloquy with Senator McClure: "I hope patience is one of my virtues, but I must say that if, in fact, the keg of dynamite is needed, you know, my fuse is lit."[116] Senator McClure told Hatfield and Evans that an agreement with the House seemed near. Hatfield and Evans did not try to force a vote on their bills, but they had made it clear that they would wait only a few weeks more for a resolution. It was being reported that Senator McClure would run for Senate Majority Leader after the November election, and some observers believed that he could not delay much longer a party colleague as influential as Senator Hatfield.[117]

The final negotiations involved such a profusion of seemingly arcane details that only the most interested professionals had the stamina to follow them. Forest Service Chief Peterson acted as a go-between, shuttling proposals back and forth to the House and Senate principals. Peterson took an optimistic approach, assuring negotiators that they were actually much closer to an agreement than they thought they were. Scott Shotwell of the National Forest Products Association thought that Peterson played an indispensable role by filling a communications vacuum.[118]

The final agreement worked out by Senator McClure and Congressmen Seiberling and Udall was much closer to the 1980 formula than to Wyoming long-term release. It did not establish any fixed time period during which the Forest Service was prohibited from reconsidering roadless land for wilderness designation but only referred to the forest plans, which were expected to last for 10 to 15 years. And during that period it did not categorically prohibit the Forest Service from managing released land so as to preserve its wilderness characteristics. It only said the agency need not do so. These two points had been the ones the environmentalists had fought most doggedly to retain from the 1980 formula. However, the timber industry and Senator McClure could claim that they had won several important concessions. The definition of a revision had been tightened up to exclude explicitly an amendment to a forest plan, thus improving the "certainty" that a forest plan would run its 10- to 15-year course before the wilderness issue had to be raised again. Also, roadless areas of less than 5,000 acres as well as those examined but found wanting for wilderness classification in Forest Service unit plans done before RARE II were also released in the first forest plans. Senator McClure was especially pleased to have won this last point because nearly 4 million acres of land in Idaho had been examined in pre-RARE II unit plans.[119]

The fight over release had come to a welcome end for all involved. The path had been cleared to designate 6.6 million acres of National Forest System land as wilderness, the largest designated acreage in a single session of Congress since the Wilderness Act of 1964. On the other hand, 13.6 million acres were "released" to regular Forest Service management.

The environmentalists' tenacity and the support they had mobilized during the wilderness leasing debate prevented the enactment of long-term release, which had seemed possible after the 1980 election. In the final stages the Forest Service played a significant role by pointing out "flaws" in the 1980 release formula. Once permanent and long-term release had died, it was helpful to have a common technical vocabulary which both sides could argue over without excessive ideological rancor.

During the summer of 1984 Congress passed 18 wilderness bills, covering 12 Eastern and 6 Western States. The most contentious and important of these was the Oregon bill, the reverberations from which have continued to affect the policies and politics of public lands in the Pacific Northwest.

CHAPTER VII

OREGON WILDERNESS

Senator Hatfield's role in resolving the release question guaranteed that the Oregon Wilderness bill would play a prominent role in 1984. But beyond its importance as a catalyst, the bill was unique because of the amount of controversy it created - not only between the timber industry and environmentalists, as was to be expected, but among the environmentalists themselves.

For the last 40 years Oregonians have fought over wilderness more passionately than the citizens of any other State. Oregon is the heart of the western timber industry, but since the 1950's it has also been the home of a growing wilderness movement that has opposed the industry's attempts to harvest the remaining pockets of the State's old-growth timber. In 1954 the Sierra Club established its first chapter outside California in Oregon. In 1960 the Oregon Cascades Conservation Council became the second statewide organization to focus exclusively on wilderness.[1] (The Montana Wilderness Coalition was first.) The movement was unsuccessful in the early years because the thriving and politically powerful timber industry parried several attempts to expand the State's wilderness system.

However, by the early 1970's the situation had begun to change. Oregon had acquired a reputation as a place of the ecologically conscious. Governor Tom McCall bolstered this image by signing the first State law banning disposable cans and by publicly discouraging migration into the State. In 1976 Robert Wazeka of the Sierra Club claimed that Eugene, OR, had become the center of the wilderness movement and was doing for that movement what Paris had done for the literary world in the 1920's, offering a "rich stimulating milieu in which ideas could be planted and then have sufficient time to grow, ferment and mature."[2]

In the 1930's the Forest Service had placed many of Oregon's high-elevation roadless areas in its primitive area system. Most of these became part of the Wilderness System when the Wilderness Act was passed in 1964. In later years wilderness activists began to move their attention down the slopes, hoping to protect lower elevation areas with more trees. The timber industry, having exhausted most of its own supply of old-growth timber, was, at the same time, trying to move up the slopes. These opposite motions brought the two groups into conflict on the middle slopes of the national forests.[3]

Both the Forest Service and Senator Mark Hatfield saw the potential for prolonged conflict over roadless areas. Since 1968, Hatfield had been in the middle of a fight over French Pete, an old-growth valley in the Willamette National Forest. In 1971 he began to explore the idea of a statewide wilderness bill to permanently settle the wilderness issue. At the same time the Forest Service was beginning its first nationwide inventory of roadless areas (RARE I). Hatfield decided to defer his plan until RARE I had been completed.[4]

The final environmental report for RARE I recommended that 262,000 acres of land be studied for possible wilderness designation. All of these acres were contiguous to already existing wilderness and were thus not "new" proposals.[5] A Sierra Club lawsuit in California resulted in an agreement by the Forest Service to prepare environmental impact statements before developing roadless areas. In Oregon the unit plan impact statements were being completed more quickly than other States because the Forest Service wanted to resolve the roadless issue in its most productive national forests. The Forest Service's unwillingness to recommend any new wilderness study areas in Oregon and the "fast tracking" of the planning process convinced some Oregon environmentalists that a new strategy was needed if a substantial number of roadless areas were to be protected.[6]

Eastern Oregon had the bulk of the State's roadless areas, but it was a virtual *terra incognita* for the Sierra Club and The Wilderness Society, which were seen as too radical by conservative Oregonians living east of the Cascades. The Northwest representatives of the Sierra Club and The Wilderness Society, Doug Scott and Joe Walicki, as well as Sierra Club activist Holly Jones, were strong believers in the value of grassroots organizing but they doubted the ability of the national organizations to make much headway in eastern Oregon. In order to circumvent this image problem, they decided to create a statewide organization that would not carry the Sierra Club stigma - an organization that would be acceptable to eastern Oregonians and would also fight vigorously to protect roadless areas.[7] Tim Mahoney gives another reason why they felt the Oregon Wilderness Coalition was needed.

> ... we believed our greatest strength in passing a wilderness bill was to have an individual constituency in every area. And Senator Hatfield was the person who had come to mind while we were working on French Pete and other areas. You would go in with your laundry list of wilderness areas and they would say: 'which of these areas is most important.' What they'd really mean is which of these areas is least important and we didn't want to answer that and we couldn't answer that. To us there is no least important area. And the way you combat that is to have a group for each area, even a small group because that group is not willing to compromise and it creates a tension which prevents somebody from paring down the wilderness list... It was designed to be difficult. We wanted it to be difficult to take an area out.[8]

As a civil engineer, Joe Walicki was the only Wilderness Society field representative hired by Clifton Merritt who did not have a background in natural resource management. But he more than made up for that lack with his ability to motivate people. During 1972 and 1973 he organized several wilderness workshops throughout Oregon. These workshops produced the people and ideas that resulted in the creation of the Oregon Wilderness Coalition (OWC) in February 1974.[9] During the first few years of its existence, the OWC was a true coalition. Its staff traveled throughout Oregon organizing local groups and instructing them in the techniques of wilderness protection and political action. It was a coordinating umbrella for all local groups of the Sierra Club, but it did not set policy. Each local group retained the right to determine how it would protect the areas it had selected. In 1976 OWC was seen as having assumed

> An active, even aggressive role in certain areas by organizing new groups; putting on statewide conferences and regional workshops focusing on techniques of wilderness preservation; monitoring and reviewing EIS's [Environmental Impact Statements] and timber sales, creating ad hoc task forces and coordinating communication, research and educational functions for Oregon wilderness activities. By contrast, its policy role has been passive. The member groups of OWC, which are fully autonomous, make decisions on specific wilderness recommendations and on strategies for implementing them.[10]

The Coalition was supported by the Sierra Club and The Wilderness Society and by contributions from its local members. OWC's first coordinator was Fred Swanson, a Sierra Club activist who taught a wilderness course at the University of Oregon and at the Survival Center in Eugene. He left in the fall of 1974 for Montana and was replaced by Jim Montieth, a soft-spoken man, who had "given up a promising scientific career as a wildlife biologist in order to fight for wilderness" at a salary of $100 a month.[11] After accepting the job, Montieth wrote a letter to OWC's Executive Council emphasizing his strong commitment to Oregon wilderness:

> In early June 1974, I spent many afternoons with Montana Fred [Swanson], for we had a lot to talk about. After about a week I told Fred I'd be willing to accept - no rather, that I wanted - the OWC Coordinator's job, if the purpose was to save all the de facto land, and more. He looked me in the eye and agreed.
>
> This is not a commitment I made to you, or even to Fred, so much as it is made to myself. This wilderness resource we deal with every day is part of us. No OWC staff person has ever felt that less than 100% of the roadless land should be retained. Regardless of so-called realities, it is this goal which will enable us to save maximum Wilderness acreage. This goal, by rational perspective, is not unrealistic or unattainable.[12]

Because of his training in wildlife biology, Montieth was instrumental in forming a statewide scientific network in support of wilderness. He was also responsible in large part for moving the debate over Oregon wilderness away from exclusive emphasis on recreational and aesthetic values. By 1976 Robert Wazeka of the Sierra Club noted that testimony on Oregon wilderness now focused much more strongly on "the need of certain wildlife species for undisturbed habitat; with concern for critical soils; and above all, with an understanding of scientific, historical and ecological values."[13]

Montieth and his staff were all in their early or middle twenties and were fired by the enthusiasms of youth. Montieth reported to his executive council that he was working 80 hours a week and was "worrying himself sick" over the fate of Oregon wilderness. He said of his field representatives Tim Lillebo and Andy Kerr that, in addition to possessing the necessary "analytical prowess", they also felt the wilderness issue "in their guts."[14] It took this kind of commitment to go into the often hostile environment of the small mill towns of southern and eastern Oregon to organize for wilderness protection.

For several years after RARE I OWC claimed it had stalled the development of roadless areas but members felt their delaying tactics eventually would fail unless more drastic action were taken. By 1977 Jim Montieth began to believe that only a national wilderness lawsuit could save Oregon roadless areas.

We believe that the only way we can insure that the majority of de facto wilderness will have even a remote chance of remaining wild is to meet the problem nationally (or at least regionally) in the courts. To date, land use plans, especially the treatment of undeveloped areas, have been totally inadequate. We feel that our legal case against land use planning will never be stronger...

Without a nationwide suit which legally challenges the process being used to destroy wilderness, conservationists stand to lose most roadless areas to development. Given our limited resources, if bad decisions are not challenged in the courts regionally or nationally, we have no method to stop the destruction of perhaps 90% of the wild lands. National court action, like the 1972 suit, will give us the de facto wilderness to work with in five years. Where would we be now if the '72 suit hadn't been filed?

The tragedy of the situation is that if we could hang onto the undeveloped areas for ten years, they would be ours. Even the timber industry admits that time is on the side of wilderness preservation.[15]

After its 1972 lawsuit the Sierra Club had become increasingly wary of another national lawsuit because of fears it would provoke a timber industry and congressional backlash. Montieth's advocacy of a lawsuit began to open a rift between the Sierra Club and the OWC staff. OWC's confrontational approach and sometimes abrasive lobbying tactics had made Montieth a controversial figure in Oregon. The OWC's Executive Council, made up primarily of Sierra Club members, was becoming disturbed over the direction in which the staff seemed to be taking the Coalition. There were allegations that the office was not running smoothly and that the staff were overstepping the bounds of their authority by attempting to set policy. The staff members, on the other hand, often chafed at their inability to make policy because they often saw problems coming before they were evident to the local groups. They also believed the Executive Council was too conservative and too willing to compromise. Montieth later said that the "problems which developed in the 1980's stemmed from the period 1976 to 1978. It was a shock to them there was so much wild land out there."[16]

In 1978, in midst of RARE II, the Executive Council attempted to oust Montieth but was unable to persuade a majority on the Governing Council. Montieth prevailed in a 2 to 1 vote. The animosity created by this split, which was essentially one between the Sierra Club and the OWC staff, persisted and colored future relations between these two organizations. The Sierra Club leadership in Oregon thought that OWC was politically naive, while OWC prided itself on its Indian and sportsmen constituency and felt that the Sierra Club was an "elitist western Oregon recreation group" populated by the "wine and brie set."[17]

OWC was convinced they were charting a new course for Oregon wilderness, one that avoided "left-wing urban anti-hunting groups" and sought out new constituencies to bring the wilderness movement more in line with basic western lifestyles. They also believed they had a mission to desanctify the wilderness idea. Wilderness to them had become a more commonplace category, one that was to be used to prevent mismanagement by the Forest Service, which they believed could not be trusted to manage land sensitively under its own authority.[18] In the 1960's and early 1970's the Sierra Club and The Wilderness Society had attacked the Forest Service for adhering to what they considered to be an overly "pure" definition of suitable wilderness land. In Oregon the Sierra Club now found itself on the receiving end of this criticism when OWC alleged that the club was interested primarily in scenic recreation areas with alpine vistas.[19] As the Sierra Club saw itself cast more and more into the role of scapegoat and even villain, some of its leaders naturally began to believe they had created a Frankenstein monster when they formed OWC in 1974.

Of the 3.4 million acres of land studied in Oregon during RARE II, the Forest Service recommended that 370,000 acres be considered for wilderness designation. The Sierra Club and OWC were deeply disappointed over this figure. OWC also charged that the agency had failed to study another million acres of land that should have been part of RARE II.[20] The environmentalists blamed these results on the agency's alleged timber bias and on the personal actions of Regional Forester Richard Worthington. When he retired in 1982, Worthington expressed his strong disapproval of the tactics of the wilderness movement. Implicit in this farewell statement was a rebuttal of OWC's argument that the only alternative to wilderness was Forest Service mismanagement:

> *Many wilderness demands are entirely unethical. They conflict with any sound, esthetic or biological value; yet they are trumpeted as being in the public interest. In reality classification of land as wilderness means locking up large acreages for the use of very few people. It is entirely contrary to basic good government when carried beyond minor amounts. The issue is not really yes or no, but how much. Most folks really don't want areas classified as wilderness. They just think they do. What they don't want is for public lands to be mistreated - they don't want poor land husbandry. The wilderness extremists have sold the public a bill of goods that the only alternative to unethical land use practices is wilderness. The cultists who advocate locking up now or lose forever are using unethical tactics to prevent unethical practices.*[21]

In 1979 Senator Hatfield introduced a 600,000-acre wilderness bill that quickly passed the Senate but stalled in the House because powerful Congressman Al Ullman (D-OR), Chairman of the House Ways and Means Committee, wanted less wilderness than Hatfield proposed while Jim Weaver wanted more. That deadlock alone would have been sufficient to block the bill but in addition the Sierra Club and The Wilderness Society opposed its permanent release provision. Angered over the failure of his bill, which he thought was a generous improvement over the RARE II recommendations, Hatfield "remained inaccessible to Sierra Club lobbyists" for the next several years. In 1981 he was reported to have said that "Mr. Weaver and company. . . are not going to have an Oregon wilderness bill in this Congress."[22] This was an indication that he would no longer take the lead in formulating a statewide bill but would wait for the House to settle its differences before returning to the issue. Despite Hatfield's seeming passivity, all the principals in the struggle over Oregon wilderness knew that in the final analysis he would be the arbiter of the Oregon wilderness bill.

Over the years Senator Hatfield had shown an independent attitude towards wilderness, which was not surprising considering the economic importance of the Oregon timber industry. He had been elected to the Senate in 1966 and thus had been involved in the creation of every Oregon wilderness since the passage of the 1964 Wilderness Act. But he had not always felt comfortable with wilderness classification, proposing instead on several occasions the creation of an intermediate "backcountry" designation that would allow more management flexibility than the wilderness designation. He also leaned more towards what the environmentalists called an "anthropocentric" view of wilderness--emphasizing scenery and recreational use over biological and ecological values.[23] OWC saw that one of its principal tasks was to convince Hatfield of the necessity of wilderness classification for the protection of fisheries, watersheds, and wildlife, especially in eastern Oregon where recreational demands were not as great as in the western part of the State. They "respected and even feared" Hatfield's staff assistant Tom Imeson "because of his competence" but they knew he would give them a fair hearing.[24] The timber industry's position was that no more or perhaps very little new wilderness should be created. Governor Vic Atiyeh's proposal of 60,000 acres was as far as the industry as a whole would go.[25] That hard-line stance made it difficult to negotiate with the industry. According to Imeson:

> *The timber industry still put up a lot of resistance to a wilderness bill. For them the common denominator was that there should be very little new wilderness. Therefore, it was hard for them to get together on a bill. They could never get a unified position near 600,000 acres, although I told them that the '79 bill was 600,000 acres and the House bill [of 1983] was 1.2 million acres and that the final bill would be somewhere in between.*[26]

After the failure of Hatfield's 1979 bill, OWC was anxious to see a new bill introduced. But here it ran into a disagreement with the Washington, DC, offices of the Sierra Club and The Wilderness Society, which, for strategic reasons dealing with the release question, thought it was more important to promote a Washington or a California bill. OWC reluctantly decided to accede to the national organizations' wishes but began to become concerned that Oregon might somehow be sacrificed in the national battle over release. Tim Mahoney's statement to Andy Kerr that an Oregon bill would be "tough" reinforced that fear.[27] Throughout the debate over release, OWC maintained that it was an arcane issue that had little relevance for Oregon because the Forest Service and timber industry would quickly develop all the released roadless areas no matter what kind of statutory language was adopted.[28]

The Sierra Club had four main reasons for delaying the introduction of an Oregon bill: (1) Hatfield's position on release was unknown after his statement following the 1980 election that the issue might be reexamined; (2) the Club did not know where Democratic Congressman Les AuCoin stood on release. As a member of the House Appropriations Committee, he was powerful and had a reputation as a good coalition-builder but he had cosponsored the 1979 national wilderness bill and had not yet endorsed the Colorado release language; (3) Congressmen AuCoin and Weaver "did not work real well together"; and finally (4) most of the proposed wilderness areas were in the district of the new Republican Congressman Denny Smith, who strenuously opposed them.[29] According to Tim Mahoney:

We were cautious about pushing an Oregon bill because we were not sure where AuCoin and Hatfield stood on release. We only wanted to work on an Oregon bill when we got a signal from Hatfield or when AuCoin indicated he understood release and would work with Weaver... The eleventh commandment of the wilderness movement is that you don't speak ill of someone else's wilderness or get in the way of their efforts to protect it. For instance, you don't make a deal for Oregon that hurts Montana.[30]

Congressman Weaver was OWC's strongest supporter in the Oregon delegation. According to his former staff assistant, Greg Skillman, he had an "almost metaphysical" belief in the value of wilderness as a genetic preserve.[31] He had opposed Senator Hatfield over the French Pete area after his election to Congress in 1974 and in 1978 had engaged in long and difficult negotiations with the senator over the Kalmiopsis area in the conference committee for the Endangered American Wilderness bill. They saw the wilderness issue differently and did not have a close working relationship. Congressman Seiberling respected Weaver for his strong advocacy of wilderness because he represented a district where the timber industry was very strong.[32] Until the 1982 election, Weaver had been opposed in the campaigns by lumbermen from the southern part of his district and in 1978 had been targeted by the Republican National Committee as the third most important Democratic congressman to defeat. According to Skillman, his boss "had stretched his district to the limit" on the wilderness issue.[33] OWC considered Weaver to be their standard bearer but even that relationship was occasionally strained when Weaver did not act as OWC thought he should. In late 1980 Weaver spoke at a political fundraiser attended by OWC and Sierra Club members. A Sierra Club member reported on the meeting and its possible consequences.

Unfortunately, Jim got carried away with the euphoria and gave wilderness a 'blank check' for the next session. It did not pass unnoticed, as several people have commented about it...

Though I doubt if we will feel repercussions too soon, it is something that we should be aware of. Since when we do reopen negotiations with OWC, we're bound to have Jim quoted back to us. I talked to Greg [Skilman] about it this evening, ... he sees ways around it, i.e. Jim meant 'legitimate' wilderness proposals, and that 'we always have to deal with Mark Hatfield.' However, I can also possibly see the SC [Sierra Club] being used as a scapegoat, if we continue to clash with OWC over the viability of some of their proposals.[34]

In 1981 Weaver on his own initiative introduced a bill capping the total amount of New Oregon wilderness at a figure that would not have subtracted more than 3 percent from the national forests' total allowable cut. One of the purposes of this bill was supposedly to show the relatively small impact that wilderness withdrawals actually would have on the Oregon economy. Weaver did not want to pass the bill but intended to use it as a vehicle to introduce a bill based on the results of these hearings. The Sierra Club and OWC drew up a list of wilderness areas containing 1.9 million acres that fell within the limit.[35] OWC had raised its hopes in 1981.

The bill is ready to move for several reasons... First, hesitation about its introduction has essentially evaporated. Key players in Congress, such as Representative John Seiberling and Senator Mark Hatfield, are flashing signals that now is the time. The Oregon bill is now a priority for national conservation organizations, who previously had felt that a Washington State bill (or others) should be moved first. The pending death of the timber industry's national release bill has contributed to the 'green light.' And current efforts to write 'bail-out' legislation for segments of the timber industry because poor market conditions have produced a political situation which will allow (even require) an Oregon Wilderness bill to begin its journey through the House of Representatives.[36]

The failure of this optimistic forecast to come to pass in the regular 1982 session of Congress made OWC impatient and fearful that more delays would result in the loss of many roadless areas.

The delays were also causing problems for the Forest Service, whose allowable timber cut was based on all the lands not recommended for wilderness in RARE II. The Forest Service had delayed cutting in controversial areas, but this restriction posed the danger of overcutting on other parts of the national forests. The agency very much wanted to have a bill to rescue it from this dilemma and to curtail the many appeals, which were hampering its activities.[37]

For the first time since his election to Congress in 1974, Weaver was opposed by a moderate Republican from Eugene, and OWC briefly considered supporting the Republican candidate out of frustration at not getting a wilderness bill from Weaver.[38] This frustration worked both ways, according to Greg Skillman: "Weaver saw himself as taking all the risks. He could never get OWC to level on numbers. The Sierra Club was more willing to compromise and prioritize."[39] OWC eventually supported Weaver in his successful re-election campaign.

On October 22 the District Court sustained Judge Karlton's 1979 decision that RARE II was inadequate, which increased the likelihood that many timber sales on Oregon's National Forests would be blocked. By this time, Congressman AuCoin, mindful of the potential danger to Oregon's economy, had promised to support Colorado release. After the election he and Weaver introduced a million-acre bill in the lameduck session. AuCoin had devised the idea of linking the wilderness bill with a contract relief section for some purchasers of federal timber who were losing money because the timber they had purchased a few years previously was too expensive to harvest.[40] Hatfield had agreed to support the bill if AuCoin could get it through the House. Hatfield planned to add contract relief in the Senate. At the last minute the National Forest Products Association decided to support the bill because of the agreement on contract relief. Congressman Denny Smith felt undercut by this decision but was still able to arouse enough opposition to prevent the bill from gathering the two-thirds majority needed under the suspension of rule procedures that had been chosen to expedite passage of the bill in the restricted time period of the lameduck session.[41] The Sierra Club had been rather pessimistic about the bill's chances, believing that AuCoin was not aware of all the obstacles in the way of passing any wilderness bill with Colorado release and burdened by the controversial contract relief provision. But its lobbyists had worked for it and were bitterly disappointed when it narrowly failed to pass. According to Andy Kerr, this defeat further widened the breach between OWC and the Sierra Club because "Doug Scott likes to be with a big winner" and feared bad consequences would follow from the bill's defeat.[42]

Having already absorbed this share of criticism for having introduced an Oregon wilderness bill, Oregon's Democratic congressmen, AuCoin, Weaver, and Ron Wyden, wanted to pass a new bill as soon as possible in the 1983 session. Oregon had been redistricted and had gained a new Republican Congressman, Bob Smith, who represented much of the area originally held by Denny Smith whose district had been moved westward. Bob Smith was an experienced state politician but during his campaign he had run on a platform of no more wilderness, which meant he too "was cut out of the process."[43] In April the House passed a 1.2-million-acre bill, 200,000 more acres having been added when the bill was in the House Public Lands Subcommittee. At first Senator Hatfield wanted to proceed with a Senate bill but then decided Oregonians deserved "another shot" at public hearings because the issue was so controversial. During the summer of 1983 Hatfield and his staff, especially Tom Imeson, held hearings and talked with all the major interests involved. He worked closely with local Forest Service personnel who responded to his many "what if" questions about boundaries and resource values.[44] Imeson visited virtually all of the proposed wilderness areas.[45]

Areas arousing concern and controversy included the North Fork John Day in eastern Oregon, Drift Creek on the Oregon Coast, Hardesty Mountain 25 miles southwest of Eugene, Middle Santiam in the Central Cascades, Waldo Lake about 60 miles east of Eugene, and Boulder Creek in southern Oregon.

OWC, which recently had renamed itself the Oregon Natural Resource Council (ONRC), had originally wanted 400,000 acres of wilderness in the North Fork John Day. This was an important fisheries area in eastern Oregon and was strongly supported by Indian tribes. The House bill had halved their proposal.

The John Day was believed to contain important mineral deposits and in the years since 1979 several large politically sophisticated companies had become interested in its potential. Eastern Oregon was economically depressed, and Hatfield wanted to protect the region's economic future.[46] Fisheries were also important to the region. At first Hatfield and his staff proposed to designate as wilderness a relatively small core area of 40,000 acres adjacent to the John Day River and place another 95,000 acres in a special fisheries management zone. In November 1983 Jean Durning, The Wilderness Society's new Northwest representative, reported on a meeting she had with Tom Imeson concerning the area:

Tom is convinced of the necessity to save its wild fish runs. But he believes that these can be saved even if occasionally selective timber harvest is allowed, perhaps with no roads. The amount of timber available probably is less important than the perception of northeast Oregonians that nobody listens to them and that a massive amount of their land is being locked away from them... The conflict on the North Fork John Day is not a recreation conflict. It is a timber (or perception) versus fish.[47]

That compromise pleased neither the environmentalists nor the timber industry. Eventually 120,000 acres were put into wilderness. (Imeson credits ONRC and The Wilderness Society as the groups primarily responsible for the creation of this important wilderness area.) The Elkhorn Mountains, part of ONRC's original John Day proposal, were dropped because, according to Tom Imeson, they were not essential for the protection of fish.[48] Their loss distressed ONRC.

The Middle Santiam is a 30,000-acre valley of low-elevation, old-growth timber on the Willamette National Forest - one of a handful still remaining in Oregon. It was heavily roaded on its periphery and had some clearcut patches inside it.[49] Its location, in traditionally Democratic Linn County, part of Weaver's district, made its fate a sensitive issue for him.[50] The House bill had put 25,000 acres into wilderness. Congressman Weaver maintained that wilderness designation for the Middle Santiam would not harm the county's economy because its mills had enough supply to last for several years and could eventually expand their radius of operations when that supply ran out. Women for Timber and other local organizations hotly disputed that contention, claiming that at least one big mill would close if the Middle Santiam were "locked up." They pointed to gravel roads and a few clearcuts within the area as evidence that it did not qualify for wilderness designation.

The area is not typical of what the majority of wilderness users want. Brush under the second growth and rotting logs under the old growth make the area virtually impassable except by established trail. This is not a high alpine fir type that wilderness users seek, but instead good, productive Forest.

... Most disturbing to people of these communities is that their own representative in Congress has stated there will be no job loss. In the late 1970's concern about the amount of available timber supply caused companies in the area to bid prices for Forest Service timber sales to unprecedented levels. Only the economic recession has given a brief respite to the competition for this timber. If the Oregon Wilderness Bill is enacted as proposed, it's only a matter of time that competition for available timber forces one large mill or two smaller ones out of business.[51]

The area also became a sore point between the Sierra Club and ONRC and illustrated their contrasting political styles. The Sierra Club's volunteer Oregon chairman, Ron Eber, felt that Hatfield was serious about his opposition to Middle Santiam and would stiffen his resistance if confronted with pressure tactics. Tom Imeson asked the Forest Service for high, middle, and low wilderness options.[52] When Imeson told Eber that Hatfield might be willing to talk about the area, Eber believed he had been offered perhaps the only opportunity he would get to include at least some of the Middle Santiam. He returned to Oregon and relayed this message to the group lobbying for the area.[53] Its leader indicated his group was flexible on boundaries. When ONRC heard about these discussions they charged that Eber had unnecessarily compromised the Middle Santiam and that no deal should have been made before the House members had negotiated with Hatfield.[54]

Several years earlier, Jim Montieth had been deeply impressed by a brief meeting with Congressman Al Ullman over a land-use issue in eastern Oregon. Ullman had told him on that occasion: "Don't you know the rules of the game? I have to move your line." Montieth drew from that encounter the lesson not to compromise until absolutely necessary, which was usually at the end of the negotiating process.[55]

Jean Durning was also deeply involved in negotiations over the Oregon bill, but she sometimes felt that she was forced to spend more time mediating between ONRC and the Sierra Club than she did working with the congressional delegation.[56]

Tom Imeson felt it had been possible to reach a compromise on Middle Santiam (eventually 7,500 acres were included) because it was only a "symbolic issue"; only Weaver among the delegation members felt deeply about it. Imeson contends it was only the Sierra Club's prodding and willingness to compromise that resulted in the inclusion of some of the area.[57] Of course, it had been more than a symbolic issue to ONRC, which for years had been talking about the need to protect as much old growth as possible, and they deeply regretted the loss of most of the Middle Santiam.

Hardesty Mountain is a popular recreation area within a 45-minute drive of Eugene. It is in the midst of a heavily developed region, most of the surrounding tracts of land having been logged and roaded. To several Sierra Club members, such as Holly Jones, Hardesty was not of wilderness quality and was seen as more of a "bargaining chip" than a legitimate proposal.[58] Greg Skillman of Weaver's staff once told some ONRC representatives that if they wanted Hardesty, they should also on the same grounds ask for Spencer's Butte, a hill in the middle of the city of Eugene. The ONRC members did not appreciate his sarcasm and from then on "they didn't like me so much."[59] Hardesty did, however, have many supporters. In the final negotiations over the bill, Hatfield offered Weaver a choice of either Hardesty or Waldo Lake and Weaver picked the larger and higher quality Waldo Lake. After the passage of the bill, Hardesty was on local front pages after an anonymous group calling itself the "Hardesty Avengers" drove spikes into trees in an attempt to prevent logging.

Drift Creek was another low-elevation old-growth area that at one time divided ONRC and the Sierra Club. It was the only potential wilderness area located in AuCoin's district. He opposed its inclusion, which made him vulnerable to charges from Bob and Denny Smith that he was willing to create wildernesses everywhere but in his own district. In 1980 the Sierra Club asked one of its members to evaluate Drift Creek's wilderness potential. While driving to the area, he looked on a map and discovered he was already in it. Shocked by this discovery, he concluded that Drift Creek did not have a core large enough to qualify it for wilderness designation.[60] But like Hardesty Mountain, it had a large number of supporters, eventually including the Sierra Club, and was backed by the Portland newspapers. Drift Creek had not been in the House bill. Hatfield wanted his proposal to be unique in some way. Hatfield felt Drift Creek had enough support to overcome AuCoin's objections and thus it became the only area in the Oregon Wilderness Act that had not been in the 1983 House bill.[61]

Until the Oregon Wilderness Act was passed, the Umpqua National Forest in southern Oregon did not have any wilderness areas. It then got three, including Boulder Creek, which had generated 22,000 responses during RARE II, the most of any area in the country. Most of these responses favored nonwilderness and thus the Forest Service had not recommended it.[62] For several years a small group of environmentalists associated with ONRC, known as the Umpqua Wilderness Defenders, had fought for wilderness in an area heavily dependent on the timber industry. At least one of the leaders of the group had to drop out when confronted with the threat of an informal economic boycott of the family business.[63] Boulder Creek was also coveted by the timber industry and was one of the few areas that Arnold Ewing, executive vice president of the Northwest Timber Association and a wilderness recreationist, regretted was in the final bill, believing it was much more suited to be a "working forest."[64]

Boulder Creek, Drift Creek, and North Fork John Day were three of the most notable successes ONRC's brand of wilderness politics, areas without great scenic attributes but valuable for their ecological and wildlife attributes.

In early 1983 ONRC's Council of Governors had directed Jim Montieth to prepare a statewide lawsuit based on the 1980 California decision in order to stop all development in Oregon RARE II roadless areas. When the House bill passed in April, its 1.2 million acres were temporarily protected by an agreement worked out by Senator Dale Bumpers and the Department of Agriculture that delayed development activities in areas undergoing wilderness consideration by Congress. Montieth convinced his board that suing while this agreement was in effect could be misinterpreted by the public. He undoubtedly was also thinking of the national organizations' strong opposition to statewide lawsuits when he wrote to his board that:

> In order to actually file the legal complaint, we must have an obvious and blatant threat to the three million acres of roadless lands in the RARE II inventory so that we demonstrate to the public, the press and politicians an act of self-defense. We need to provide compelling evidence that our action was taken because our backs were against the wall, that we were being irreparably and irreversibly harmed on a grand scale. With the Bumpers agreement in place, none of the most popular Wilderness proposals (1.2 million acres, the ones currently in HR 1149)

are threatened. If we filed now, it is possible the public would misread our action. It is even possible, although unlikely, that a judge might dismiss our complaint as ungrounded.[65]

When Hatfield did not introduce his bill in the fall and time ran out in the 1983 session, Montieth could no longer use the now-expired Bumpers Agreement to dissuade his board. The House bill was dead until the next session. Consequently, the Bumpers Agreement was no longer in effect. With the Sierra Club's financial backing, Oregon environmentalists had filed two site-specific lawsuits. They had lost in the case of the Bald Mountain Road on the Kalmiopsis, which had not been covered by the Bumpers Agreement because it was not in the House bill in deference to Senator Hatfield, who had opposed its inclusion in the 1978 Endangered American Wilderness Act.

The direct-action environmental group known as Earth First was conducting a delaying-action blockade of the road, which was generating publicity and some money but was not stopping the development feared by ONRC. In the past ONRC had used the threat of a suit to get action from politicians and the national conservation organizations but now some were claiming that ONRC was bluffing. Andy Kerr and other ONRC staffers felt they had no alternative but to file a suit.[66]

Mike Anderson of the Umpqua Wilderness Defenders told Jim Montieth he knew of an attorney, Neil Kagan, in Roseburg who would take on the case. On December 2 ONRC decided to sue.[67] The national conservation organizations and members of the Oregon delegation were informed a day or two before the filing.

The lack of much notice angered some, especially Congressman AuCoin, who was in a close reelection fight with a timber company executive. Greg Skillman believed that "the idea of a lawsuit was a trump card. If you play it, it's over... We had a day warning about it. That was the last straw in a deteriorating relationship."[68] The Sierra Club and The Wilderness Society attempted to dissociate themselves completely from ONRC's suit. The National Audubon Society, whose local chapters in Oregon had been working closely with ONRC, joined in the suit. (Brock Evans in the Society's Washington Office was the only top official with a long background in national wilderness politics and at the time of the lawsuit he was preparing to launch a congressional campaign in Washington State.)[69] According to Andy Kerr, the Sierra Club and The Wilderness Society did a "tremendous job of damage control" by effectively distancing themselves from ONRC's actions and by convincing political leaders they were not part of a plot to undermine political negotiations.[70]

Montieth explained to his members that one of the reasons ONRC had decided to sue was because other groups were planning to file and if they did, ONRC would be "perceived as the momentum behind it." He went on to explain the political rationale behind the suit.

A RARE II suit may provide the necessary incentive, since the 'locking up' of three million acres will be anathema to them [the timber industry], and the only way to 'unlock' those lands is to pass legislation. Since the industry, for the time being, has much more timber than they can use, they may still wish to wait before settling on RARE II. The lawsuit is the best catalyst we can provide, short of total capitulation on the acreage. If it's not an adequate incentive for the industry, so be it.

. . .

Even though it doesn't include all the roadless areas in Oregon, the RARE II lawsuit will bring needed perspective to HR 1149, and alert the public to the fact that it is a compromise bill... They [the House members] should feel less pressure to compromise on 'their' 1.2 million acre bill if they remember we're not satisfied with it either.[71]

Before Hatfield unveiled his plan in February 1984 there were reports that 1 million acres was the "magic number" his bill would not exceed. At a summer field hearing he was reported to have said the bill would probably be somewhere between 600,000 and 1 million acres. At about the same time one of his staff assistants said the bill would be about 900,000 acres. However, it was Tom Imeson's assumption that there was no magic number because that was "too crass" a way to deal with the issue, which he thought should be approached based on the merits of the individual areas.[72] But ONRC concluded from conversations with congressional staffers that an agreement had been reached not to exceed 1 million acres.[73] They began to dig in their heels against a reduction of what they thought was already a modest 1.2-million-acre House bill.

When Hatfield told the Sierra Club leaders at his February meeting that he would introduce a bill close to 1 million acres, Ron Eber was "astonished" because he thought it would be much lower. (His enthusiasm was tempered when the Senator included a lot of "backcountry" in the bill.)[74] In the fall of 1983 Tim Mahoney had told Hatfield's staff that the Sierra Club would not press for a conference committee between the House and Senate if the Senator's bill was at least "moderately good." The 950,000-acre bill, including the backcountry Oregon Cascades Recreation Area and John Day Fish Management Zone (later changed to wilderness), lived up to that description. Moreover, the Sierra Club was very pleased with Hatfield's role in breaking the deadlock over release language. They stuck by their agreement, which effectively excluded them from most of the final negotiations.[75]

Hatfield stated that as chairman of the Senate Appropriations Committee he would have no time for a conference committee. He agreed to listen to requests from the House but asserted that his bill would have to be accepted essentially as it was if the delegation wanted a wilderness act before the 1984 election. The Wilderness Society and the Sierra Club took him at his word, as did Congressmen AuCoin and Wyden who accepted his bill, which brought charges from ONRC that they had "capitulated." (They said they wished they had gotten 1.2 million acres but that was "just not to be.")[76]

The bill's final months were painful for the ONRC staff. For years they had proclaimed their goal of protecting all of the 3.4 million acres of RARE II roadless land. They had recruited local groups to support almost all of these areas and thus it was impossible for them to engage in political negotiations that appeared to be "selling them out." The Sierra Club, which was more willing to "prioritize", felt that ONRC was "setting up their members for a big loss" that they could then blame on the Club and Congressmen AuCoin and Wyden.[77] ONRC had wanted a conference committee which they thought might have increased the acreage by as much as 100,000 to 150,000 acres.[78] The Sierra Club was very anxious to avoid a conference because they were afraid that Senator McClure would be involved.[79] Ron Eber believes that even if a conference had been held it would have resulted in an increase of only about 30,000, "which certainly cannot be considered a capitulation."[80]

Congressman Weaver had not accepted Hatfield's bill and at the last minute was able to get two final changes - the addition of the rest of the Grassy Knob area and the substitution of Monument Rock for what was believed to be the less threatened Glacier Peak area. Looking back on the tortured history of the bill, Tim Mahoney believes that all the participants played essential roles in its passage:

Seiberling was the go-between at the end. The delegation was not getting along. Hatfield made two last minute changes, adding Grassy Knob in Weaver's District, and switching Monument for Glacier after Weaver asked ONRC to choose, the only time in the whole process where ONRC had made a choice. It seemed each person or group believes he was the key in the final resolution. AuCoin believes if he hadn't worked with Hatfield in that earlier period, Hatfield would not have gone for soft release and if he had not endorsed the Hatfield bill, it would have fallen apart at the last minute. Weaver believes that if AuCoin had been more hard line we could have had a better bill. If Weaver had not hung out in the end, we could not have gotten Monument or Grassy Knob and he's right and AuCoin is right. ONRC believes that if the Sierra Club had been more confrontational we could have had a bigger bill and if we had called for a conference we could have jerked Hatfield up a little further. They believe because they didn't give up, they got Grassy Knob and Monument Rock and maybe they could have had others. They're probably right. But we believe we never would have reached that point if we hadn't worked the bill the way we did. Despite all the in-fighting that took place in Oregon within the delegation and the environmental community, it worked in almost a textbook way, having both inside and outside players and hardline and softline people who were able to get the bill through. The hardliners were able to improve it over what the softliners could get and the softliners were able to keep the machinery running.[81]

Chapter VIII
An End, and a New Beginning

The Wilderness Act has had a profound effect upon the Forest Service - the federal agency most involved in the formulation of the act's basic principles. Because the act involved fundamental questions of definition, its passage was a welcome opportunity to clarify policy regarding uses of national forest wilderness areas.

Casualties of language litter the battlefields of public policy, and in the case of the wilderness bill, the expression "multiple use" (in foresters' parlance, that combination of uses yielding the greatest public benefit) came close to linguistic extinction. Because it had been coined by practical foresters interested in using commercial forest resources, the term usually put environmentalists on the defensive. When they did care to pronounce it, they meant only those uses that were consistent with maintaining the forest's wilderness character - scientific research, recreational and scenic enjoyment, educational enrichment, and watershed protection. In fact, Howard Zahniser maintained that watershed protection was the dominant use of any wilderness area; if recreationists threatened this purpose, their numbers and activities should be restricted.[1] The term was rescued from a complete semantic muddle only with the passage of the Multiple-Use, Sustained-Yield Act of 1960, which recognized that "the establishment and maintenance of areas of wilderness are consistent with the purposes and provisions" of the act.[2]

In the national forests, the debate over multiple use and wilderness restrictions had focused on timber cutting and grazing. The Wilderness Act clarified Forest Service policy in both areas, although prescribing different approaches to each activity. A wilderness bill that allowed timber harvesting would have been no wilderness bill at all, for it would have licensed the largest possible alteration of wilderness environments. Some primitive areas had been partially harvested under the L-20 Regulations, but commercial timbering was completely prohibited under U Regulations and the Wilderness Act, except that wilderness miners could cut timber for their operations if it was not otherwise available and if the cutting was done under good forest management practices defined under national forest rules and regulations.[3]

Grazing restrictions were somewhat more complex. A reading of the statements of the bill's sponsors, drafters, and supporters over the years yields several general conclusions about the intent of the language on grazing in the Wilderness Act. First, the secretary of agriculture can regulate established wilderness grazing but not abolish it. Grazing is a permanent, legitimate use of national forest wilderness. However, the drafters intended to freeze the status quo in wilderness grazing. For instance, grazing cannot be introduced into a wilderness area that has not previously known it. Structures and improvements used for grazing can be maintained, but no new facilities can be built unless, like drift fences, they are needed to protect the wilderness environment. Second, the Forest Service can promulgate reasonable grazing regulations, as it does in other national forest areas. It can, for example, reduce stocking in areas that it finds to be overgrazed. Third, graziers have not acquired "vested rights" in their wilderness permit areas; grazing in wilderness areas is a "privilege" just as it is elsewhere in the national forests.[4]

Zahniser once said that wilderness, like chastity, is defined by that which it negates.[5] The history of the wilderness bill is essentially the history of how much the environmentalists could compromise before they themselves were compromised. When it was all over, some felt that they had given too much.[6] It is true that the preservationists conceded several important points in order to placate other interest groups and move the bill out of congressional committees. But as a soap opera character might say, since the preservationists only accepted the violation of their principles under duress, they were never really violated. More important, those who championed the wilderness concept, from Aldo Leopold to Howard Zahniser, created a strong popular movement that has been able to vitiate some of what its advocates perceived as compromising provisions.[7] While creating legislative sanction for the wilderness system, they also generated tremendous public support for their cause. Consequently, it has been possible to preserve and enlarge the wilderness system despite what once seemed to be disabling provisions in the act. Thus, the evolution of the wilderness concept did not stop with the signing of the Wilderness Act; it has been clarified and expanded by subsequent legislation.

Except for Montana and Idaho, which at the time of the writing of this conclusion (1994) still have not gotten state wilderness acts, the 1984 wilderness legislation marked the beginning of the end of an era of large additions to the national forest portion of the Wilderness System. In coming years the System will undoubtedly grow, especially in Alaska, Montana, and Idaho, but it will do so at a slower rate than it did from 1964 to 1984.

With the perspective of fifteen years of hindsight, it seems clear that RARE I and RARE II, once maligned by environmentalists, actually greatly benefited their cause.[8] These national studies identified millions of acres of roadless land the environmentalists had not known about, created grassroots support for their cause, and put pressure on Congress to resolve the impasse over wilderness and resource development. But the post-RARE II era also has its drawbacks. In some instances, especially in Oregon, national environmental organizations clashed with local wilderness movement groups revealing for the first time in the history of the wilderness movement a serious internal strain. Also, unlike the pre-RARE II era when environmentalists almost always won individual battles on terrain they had chosen to fight over, in the post-RARE II era their victories in statewide legislation were usually bittersweet. Even the jubilation over the Washington Wilderness Act, which created more than a million acres of wilderness in a State where the Forest Service had recommended 270,00 acres was tempered by the fact that the highly valued Kettle Range was left out.[9] In some cases roadless areas became poker chips that were moved in and out of the Wilderness System without much regard for their intrinsic merits.[10]

In the Pacific Northwest the wilderness issue did not fade away after what the environmentalists considered to be the disappointments of the Oregon and, to a lesser extent, the Washington Wilderness Acts but metamorphosed into the struggle over "old growth", "Ancient Forests", and the northern spotted owl. As Steven Lewis Yaffee points out in his comprehensive study of the fight over the spotted owl, the drive to save owls was more than just an attempt to get wilderness by other means. Nevertheless, the wilderness organizations were certainly among the most important participants in it.[11] The desire to exclude timber harvesting from low-elevation "old growth", the principal habitat of the owl, was a logical progression from the earlier effort to preserve as wilderness the ecological integrity of mid-elevation forests.

Another result of the years of wrangling over wilderness has been a gradual but profound change in the Forest Service. Once an agency dominated by foresters operating according to a "multiple-use" ethic that placed timber above all other uses, the Forest Service's professions and multiple-uses now relate on a much more democratic footing. This change was epitomized in December of 1993 when Jack Ward Thomas, a leading Forest Service wildlife biologist and head of the Interagency Scientific Committee on the spotted owl, was appointed chief of the Forest Service.

Appendix A - National Wilderness Preservation System:*

Agency	Units	Federal Acres	(%)	Hectares
Forest Service, USDA	398	34,570,227	(36.0)	13,990,379
National Park Service, USDI	42	39,140,023	(40.7)	15,839,750
Fish and Wildlife Service, USDI	75	20,676,341	(21.5)	8,367,600
Bureau of Land Management, USDI	68	1,669,522	(1.8)	38,873,376
GRAND TOTAL	**564**	**96,056,113**	**(100)**	**38,873,376**

National Wilderness Preservation System (excluding Alaska):

Agency	Units	Federal Acres	(%)	Hectares
Forest Service, USDA	379	28,817,121	(74.6)	11,662,129
National Park Service, USDI	34	6,160,653	(15.9)	2,493,182
Fish and Wildlife Service, USDI	54	2,00,021	(5.2)	809,397
Bureau of Land Management, USDI	68	1,669,522	(4.3)	675,646
TOTAL	**516**	**38,647,317**	**(100)**	**15,640,354**

National Wilderness Preservation System (Alaska):

Agency	Units	Federal Acres	(%)	Hectares
Forest Service, USDA	19	5,753,106	(10.0)	2,328,250
National Park Service, USDI	8	32,979,370	(57.4)	13,346,568
Fish and Wildlife Service, USDI	21	18,676,320	(32.5)	7,558,203
TOTAL	**48**	**57,408,796**	**(100)**	**23,233,021**

Note: Detailed breakdowns by wilderness within each State and Agency jurisdiction can be found in the Annual Wilderness Report to Congress. Some acreage values are estimates, pending final mapping and surveys. Total number of units for all agencies is 564; this is not additive from information above because of overlapping responsibilities. Date prepared: 9/1/93.

* The Annual Report to Congress, mandated by the Wilderness Act, describes the current year's National Wilderness Preservation System (NWPS) statues for the Forest Service.

Appendix B
Designated Wilderness Areas, Forest Service, 1993

Wilderness Area Name	Administrative Unit Name(s)	Size in Acres (Federal only)	Acres Inholding	Public Law Number	Date(s) of Designation
Geographic State: Alaska					
Chuck River	Tongass NF	74,278	692	101-626	11/28/90
Coronation Island	Tongass NF	19,232		96-487	12/20/80
Endicott River	Tongass NF	98,729		96-487	12/20/80
Karta River	Tongass NF	39,889	5	101-626	11/28/90
Kootznoowoo - Admiralty Island	Tongass NF	955,921	32,129	96-487	12/20/80
Kootznoowoo - Young Lake Addition	Tongass NF			101-626	11/28/90
Kuiu	Tongass NF	60,581		101-626	11/28/90
Maurelle Islands	Tongass NF	4,937		96-487	12/20/80
Misty Fjords	Tongass NF	2,142,243	664	96-487	12/20/80
Petersburg Creek-Duncan Salt Chuck	Tongass NF	46,777	72	96-487	12/20/80
Pleasant/Lemusurier/Inian Islands	Tongass NF	23,096	55	101-626	11/28/90
Russell Fjord	Tongass NF	348,701		96-487	12/20/80
South Baranof	Tongass NF	319,568		96-487	12/20/80
South Etolin	Tongass NF	83,371		101-626	11/28/90
South Prince of Wales	Tongass NF	90,996	22	96-487	12/20/80
Stikine-LeConte	Tongass NF	448,841	1,110	96-487	12/20/80
Tebenkof Bay	Tongass NF	66,839		96-487	12/20/80
Tracy Arm-Fords Terror	Tongass NF	653,179		96-487	12/20/80
Warren Island	Tongass NF	11,181		96-487	12/20/80
West Chichagof-Yakobi	Tongass NF	264,747	782	96-487	12/20/80
Total Forest Service acres in Alaska:		**5,753,106**	**35,531**		
Geographic State: Alabama					
Cheaha	Talladega NF	7,245		97-411	01/03/83
Cheaha	Talladega NF			100-547	10/28/88
Sipsey	William B. Bankhead NF	25,906	80	93-622	01/03/75
Sipsey	William B. Bankhead NF			100-547	10/28/88
Total Forest Service acres in Alabama:		**33,151**	**80**		
Geographic State: Arkansas					
Black Fork Mountain	Ouachita NF	8,350	80	98-508	10/19/84
Caney Creek	Ouachita NF	14,460		93-622	01/03/75
Dry Creek	Ouachita NF	6,310		98-508	10/19/84
East Fork	Ozark NF	10,688		98-508	10/19/84
Flatside	Ouachita NF	9,507		98-508	10/19/84
Hurricane Creek	Ozark NF	15,307	120	98-508	10/19/84
Leatherwood	Ozark NF	16,838	142	98-508	10/19/84
Poteau Mountain	Ouachita NF	11,299		98-508	10/19/84
Richland Creek	Ozark NF	11,801		98-508	10/19/84
Upper Buffalo	Ozark NF	12,000	35	93-622	01/03/75
Total Forest Service acres in Arkansas:		**116,560**	**377**		
Geographic State: Arizona					
Apache Creek	Prescott NF	5,666		98-406	08/28/84
Bear Wallow	Apache NF	11,080		98-406	08/28/84
Castle Creek	Prescott NF	25,215		98-406	08/28/84
Cedar Bench	Prescott NF	14,950		98-406	08/28/84
Chiricahua	Coronado NF	87,700		88-577	09/03/64
Chiricahua	Coronado NF			98-406	08/28/84
Escudilla	Apache NF	5,200		98-406	08/28/84
Fossil Springs	Coconino NF	22,149		98-406	08/28/84
Four Peaks	Tonto NF	61,074		98-406	08/28/84
Galiuro	Coronado NF	76,317		88-577	09/03/64
Galiuro	Coronado NF			98-406	08/28/84
Granite Mountain	Prescott NF	9,762		98-406	08/28/84
Hellsgate	Tonto NF	37,440		98-406	08/28/84
Juniper Mesa	Prescott NF	7,406		98-406	08/28/84
Kachina Peaks	Coconino NF	18,616		98-406	08/28/84
Kanab Creek	Kaibab NF	63,760		98-406	08/28/84
Kendrick Mountain	Coconino NF	1,510		98-406	08/28/84
Kendrick Mountain	Kaibab NF	5,000		98-406	08/28/84
Mazatzal	Coconino NF	4,275		98-406	08/28/84
Mazatzal	Tonto NF	248,115	104	88-577	09/03/64
Mazatzal	Tonto NF			98-406	08/28/84
Miller Peak	Coronado NF	20,190		98-406	08/28/84
Mount Baldy	Apache NF	7,079		91-504	10/23/70
Mount Wrightson	Coronado NF	25,260		98-406	08/28/84
Munds Mountain	Coconino NF	24,411		98-406	08/28/84
Pajarita	Coronado NF	7,553		98-406	08/28/84
Pine Mountain	Prescott NF	8,609		92-230	02/15/72
Pine Mountain	Tonto NF	11,452		92-230	02/15/72
Punch Ridge	Coronado NF	56,933		95-237	02/24/78
Red Rock-Secret Mountain	Coconino NF	47,194		98-406	08/28/84
Rincon Mountain	Coronado NF	38,590		98-406	08/28/84
Saddle Mountain	Kaibab NF	40,539		98-406	08/28/84
Salome	Tonto NF	18,531		98-406	08/28/84
Salt River Canyon	Tonto NF	32,101		98-406	08/28/84
Santa Teresa	Coronado NF	26,780		98-406	08/28/84
Sierra Ancha	Tonto NF	20,850		88-577	09/03/64
Strawberry Crater	Coconino NF	10,743		98-406	08/28/84
Superstition	Tonto NF	159,757	23	88-577	09/03/64
Superstition	Tonto NF			98-406	08/28/84
Sycamore Canyon	Coconino NF	23,325	5	92-241	03/06/72
Sycamore Canyon	Coconino NF			98-406	08/28/84
Sycamore Canyon	Kaibab NF	7,125		92-241	03/06/72

Appendix B
Designated Wilderness Areas, Forest Service, 1993

Wilderness Area Name	Administrative Unit Name(s)	Size in Acres (Federal only)	Acres Inholding	Public Law Number	Date(s) of Designation
Sycamore Canyon	Prescott NF	25,487		92-241	03/06/72
Sycamore Canyon	Prescott NF			98-406	08/28/84
West Clear Creek	Coconino NF	15,238		98-406	08/28/84
Wet Beaver	Coconino NF	6,155		98-406	08/28/84
Woodchute	Prescott NF	5,833		98-406	08/28/84
Total Forest Service acres in Arizona:		**1,344,970**	**132**		
Geographic State: California					
Aqua Tibia	Cleveland NF	15,933		93-632	01/03/75
Ansel Adams	Inyo NF	78,775	2	88-577	09/03/64
Ansel Adams	Sierra NF	151,483		88-577	09/03/64
Ansel Adams	Sierra NF			98-425	09/28/84
Bucks Lake	Plumas NF	21,000		98-425	09/28/84
Caribou	Lassen NF	20,546		88-577	09/03/64
Caribou	Lassen NF			98-425	09/28/84
Carson-Iceberg	Stanislaus NF	77,993	320	98-425	09/28/84
Carson-Iceberg	Toiyabe NF	83,188		98-425	09/28/84
Castle Crags	Shasta NF	8,627	2,421	98-425	09/28/84
Chanchelulla	Trinity NF	8,200		98-425	09/28/84
Chumash	Los Padres NF	38,150	50	102-301	06/19/92
Cucamonga	Angeles NF	4,200		98-425	09/28/84
Cucamonga	San Bernardino NF	8,581		88-577	09/03/64
Desolation	Eldorado NF	63,475		91-82	10/10/69
Dick Smith	Los Padres NF	67,800	200	98-425	09/28/84
Dinkey Lakes	Sierra NF	30,000		98-425	09/28/84
Dome Land	Sequoia NF	93,781	80	88-577	09/03/64
Dome Land	Sequoia NF			98-425	09/28/84
Emigrant	Stanislaus NF	112,277	61	93-632	01/03/75
Emigrant	Stanislaus NF			98-425	09/28/84
Garcia	Los Padres NF	14,100		102-301	06/19/92
Golden Trout	Inyo NF	192,765	1,553	95-237	02/24/78
Golden Trout	Sequoia NF	110,746	400	95-237	02/24/78
Granite Chief	Tahoe NF	19,048	6,700	98-425	09/28/84
Hauser	Cleveland NF	7,547	544	98-425	09/28/84
Hoover	Inyo NF	9,507	21	88-577	09/03/64
Hoover	Toiyabe NF	39,094		88-577	09/03/64
Ishi	Lassen NF	41,099	1,767	98-425	09/28/84
Jennie Lakes	Sequoia NF	10,289		98-425	09/28/84
John Muir	Inyo NF	228,366	178	88-577	09/03/64
John Muir	Sierra NF	351,957	642	88-577	09/03/64
John Muir	Sierra NF			98-425	09/28/84
Kaiser	Sierra NF	22,700		.-557	10/19/76
Machesna Mountain	Los Padres NF	19,760	240	98-425	09/28/84
Marble Mountain	Klamath NF	241,744	720	88-577	09/03/64
Marble Mountain	Klamath NF			98-425	09/28/84
Matilija	Los Padres NF	29,600		102-301	06/19/92
Mokelumne	Eldorado NF	60,154	1,677	88-577	09/03/64
Mokelumne	Eldorado NF			98-425	09/28/84
Mokelumne	Stanislaus NF	22,267	10	88-577	09/03/64
Mokelumne	Stanislaus NF			98-425	09/28/84
Mokelumne	Toiyabe NF	16,740		98-425	09/28/84
Monarch	Sequoia NF	24,152		98-425	09/28/84
Monarch	Sierra NF	20,744		98-425	09/28/84
Mount Shasta	Shasta NF	33,845	3,865	98-425	09/28/84
North Fork	Six Rivers NF	7,999	101	98-425	09/28/84
Pine Creek	Cleveland NF	13,480	206	98-425	09/28/84
Red Buttes	Rogue River NF	16,150		98-425	09/28/84
Russian	Klamath NF	12,000		98-425	09/28/84
San Gabriel	Angeles NF	36,118		90-318	05/24/68
San Gorgonio	San Bernardino NF	56,722	1,947	88-577	09/03/64
San Gorgonio	San Bernardino NF			98-425	09/28/84
San Jacinto	San Bernardino NF	32,248	1,160	88-577	09/03/64
San Jacinto	San Bernardino NF			98-425	09/28/84
San Mateo Canyon	Cleveland NF	38,484	1,056	98-425	09/28/84
San Rafael	Los Padres NF	150,980	190	90-271	03/21/68
San Rafael	Los Padres NF			98-425	09/28/84
San Rafael	Los Padres NF	46,400		102-301	06/19/92
Santa Lucia	Los Padres NF	18,679	3,025	95-237	02/24/78
Santa Rosa	San Bernardino NF	13,787	6,016	98-425	09/28/84
Sespe	Angeles NF			102-301	06/19/92
Sespe	Los Padres NF	219,700		102-301	06/19/92
Sheep Mountain	Angeles NF	39,482	484	98-425	09/28/84
Sheep Mountain	San Bernardino NF	2,401		98-425	09/28/84
Silver Peak	Los Padres NF	14,500		102-301	06/19/92
Siskiyou	Klamath NF	75,680	320	98-425	09/28/84
Siskiyou	Siskiyou NF	5,300		98-425	09/28/84
Siskiyou	Six Rivers NF	71,700		98-425	09/28/84
Snow Mountain	Mendocino NF	36,370	630	98-425	09/28/84
South Sierra	Inyo NF	31,865		98-425	09/28/84
South Sierra	Sequoia NF	28,219	240	98-425	09/28/84
South Warner	Modoc NF	70,614	115	88-577	09/03/64
South Warner	Modoc NF			98-425	09/28/84
Thousand Lakes	Lassen NF	16,335		88-577	09/03/64
Trinity Alps	Klamath NF	77,860	1,840	98-425	09/28/84
Trinity Alps	Shasta NF	102,821	12,179	98-425	09/28/84
Trinity Alps	Six Rivers NF	25,400		98-425	09/28/84
Trinity Alps	Trinity NF	292,060	940	98-425	09/28/84
Ventana	Los Padres NF	164,178	3,311	91-58	08/16/69
Ventana	Los Padres NF			95-237	02/24/78
Ventana	Los Padres NF			98-425	09/28/84
Ventana	Los Padres NF	38,000		102-301	06/19/92
Yolla Bolly-Middle Eel	Mendocino NF	98,323	3,230	88-577	09/03/64
Yolla Bolly-Middle Eel	Mendocino NF			98-425	09/28/84
Yolla Bolly-Middle Eel	Six Rivers NF	10,813	620	98-425	09/28/84
Yolla Bolly-Middle Eel	Trinity NF	37,560	1,080	88-577	09/03/64
Total Forest Service acres in California:		**4,302,461**	**60,141**		

Appendix B
Designated Wilderness Areas, Forest Service, 1993

Wilderness Area Name	Administrative Unit Name(s)	Size in Acres (Federal only)	Acres Inholding	Public Law Number	Date(s) of Designation
Geographic State: Colorado					
Big Blue	Uncompahgre NF	98,516	68	96-560	12/22/80
Cache La Poudre	Roosevelt NF	9,238	70	96-560	12/22/80
Collegiate Peaks	Gunnison NF	48,986	108	96-560	12/22/80
Collegiate Peaks	San Isabel NF	82,248	983	96-560	12/22/80
Collegiate Peaks	White River NF	35,482	189	96-560	12/22/80
Comanche Peak	Roosevelt NF	66,791	110	96-560	12/22/80
Eagles Nest	Arapaho NF	82,324	67	94-352	07/12/76
Eagles Nest	White River NF	50,582	523	94-352	07/12/76
Flat Tops	Routt NF	38,870		94-146	12/12/75
Flat Tops	White River NF	196,165	195	94-146	12/12/75
Holy Cross	San Isabel NF	9,489	79	96-560	12/22/80
Holy Cross	White River NF	112,899	943	96-560	12/22/80
Hunter-Fryingpan	White River NF	74,399	200	95-237	02/24/78
Indian Peaks	Arapaho NF	40,109	71	95-450	10/11/78
Indian Peaks	Arapaho NF			96-560	12/22/80
Indian Peaks	Roosevelt NF	30,265	449	95-450	10/11/78
Indian Peaks	Roosevelt NF			96-560	12/22/80
La Garita	Gunnison NF	79,822		88-577	09/03/64
La Garita	Gunnison NF			96-560	12/22/80
La Garita	Rio Grande NF	24,164		88-577	09/03/64
La Garita	Rio Grande NF			96-560	12/22/80
Lizard Head	San Juan NF	20,802	157	96-560	12/22/80
Lizard Head	Uncompahgre NF	20,391	146	96-560	12/22/80
Lost Creek	Pike NF	105,090	361	96-560	12/22/80
Maroon Bells-Snowmass	Gunnison NF	19,194	1,170	96-560	12/22/80
Maroon Bells-Snowmass	White River NF	161,768	1,715	88-577	09/03/64
Maroon Bells-Snowmass	White River NF			96-560	12/22/80
Mount Evans	Arapaho NF	40,274		96-560	12/22/80
Mount Evans	Pike NF	34,127		96-560	12/22/80
Mount Massive	San Isabel NF	27,980	67	96-560	12/22/80
Mount Sneffels	Uncompahgre NF	16,565	22	96-560	12/22/80
Mount Zirkel	Routt NF	139,818	80	88-577	09/03/64
Mount Zirkel	Routt NF			96-560	12/22/80
Neota	Roosevelt NF	9,657		96-560	12/22/80
Neota	Routt NF	267		96-560	12/22/80
Never Summer	Arapaho NF	7,098	343	96-560	12/22/80
Never Summer	Routt NF	6,659		96-560	12/22/80
Platte River	Routt NF	743		98-550	10/30/84
Raggeds	Gunnison NF	43,062	36	96-560	12/22/80
Raggeds	White River NF	16,457	375	96-560	12/22/80
Rawah	Roosevelt NF	71,606	866	88-577	09/03/64
Rawah	Roosevelt NF			96-560	12/22/80
Rawah	Routt NF	1,462		96-560	12/22/80
South San Juan	Rio Grande NF	87,847		96-560	12/22/80
South San Juan	San Juan NF	39,843		96-560	12/22/80
Weminuche	Rio Grande NF	164,715	52	93-632	01/03/75
Weminuche	Rio Grande NF			96-560	12/22/80
Weminuche	San Juan NF	294,889	4,022	93-632	01/03/75
Weminuche	San Juan NF			96-560	12/22/80
West Elk	Gunnison NF	176,172	240	88-577	09/03/64
West Elk	Gunnison NF			96-560	12/22/80
Total Forest Service acres in Colorado:		**2,586,835**	**13,707**		
Geographic State: Florida					
Alexander Springs	Ocala NF	7,941		98-430	09/28/84
Big Gum Swamp	Osceola NF	13,660		98-430	09/28/84
Billies Bay	Ocala NF	3,092		98-430	09/28/84
Bradwell Bay	Apalachicola NF	24,602		93-622	01/03/75
Bradwell Bay	Apalachicola NF			98-430	09/28/84
Juniper Prairie	Ocala NF	14,277	4	98-430	09/28/84
Little Lake George	Ocala NF	2,833		98-430	09/28/84
Mud Swamp/New River	Apalachicola NF	8,090		98-430	09/28/84
Total Forest Service acres in Florida:		**74,495**	**4**		
Geographic State: Georgia					
Big Frog	Chattahoochee NF	83		98-578	10/30/84
Blood Mountain	Chattahoochee NF	7,800		102-217	12/11/91
Brasstown	Chattahoochee NF	11,178	227	99-555	10/27/86
Brasstown	Chattahoochee NF	1,160		102-217	12/11/91
Cohutta	Chattahoochee NF	35,143	4	93-622	01/03/75
Cohutta	Chattahoochee NF			99-555	10/27/86
Ellicott Rock	Chattahoochee NF	2,181		93-622	01/03/75
Ellicott Rock	Chattahoochee NF			98-514	10/19/84
Mark Trail	Chattahoochee NF	16,400		102-217	12/11/91
Raven Cliffs	Chattahoochee NF	8,562		99-555	10/27/86
Rich Mountain	Chattahoochee NF	9,476	173	99-555	10/27/86
Southern Nantahala	Chattahoochee NF	11,770		98-514	10/19/84
Tray Mountain	Chattahoochee NF	9,702		99-555	10/27/86
Total Forest Service acres in Georgia:		**113,455**	**404**		
Geographic State: Idaho					
Frank Church-River of No Return	Bitterroot NF	193,703		96-312	07/23/80
Frank Church-River of No Return	Bitterroot NF			98-231	03/14/84
Frank Church-River of No Return	Boise NF	332,891	1,822	96-312	07/23/80
Frank Church-River of No Return	Boise NF			98-231	03/14/84
Frank Church-River of No Return	Challis NF	515,421	434	96-312	07/23/80
Frank Church-River of No Return	Challis NF			98-231	03/14/84
Frank Church-River of No Return	Nez Perce NF	110,698	213	96-312	07/23/80
Frank Church-River of No Return	Nez Perce NF			98-231	03/14/84
Frank Church-River of No Return	Payette NF	791,675	3,840	96-312	07/23/80
Frank Church-River of No Return	Payette NF			98-231	03/14/84
Frank Church-River of No Return	Salmon NF	421,433	1,201	96-312	07/23/80

Appendix B
Designated Wilderness Areas, Forest Service, 1993

Wilderness Area Name	Administrative Unit Name(s)	Size in Acres (Federal only)	Acres Inholding	Public Law Number	Date(s) of Designation
Frank Church-River of No Return	Salmon NF			98-231	03/14/84
Gospel Hump	Nez Perce NF	205,764	236	95-237	02/24/78
Hells Canyon	Nez Perce NF	59,900		94-199	12/31/75
Hells Canyon	Payette NF	23,911	289	94-199	12/31/75
Sawtooth	Boise NF	150,071		92-400	08/22/72
Sawtooth	Challis NF	12,020		92-400	08/22/72
Sawtooth	Sawtooth NF	54,997		92-400	08/22/72
Selway-Bitterroot	Bitterroot NF	270,321	53	88-577	09/03/64
Selway-Bitterroot	Bitterroot NF			96-312	07/23/80
Selway-Bitterroot	Clearwater NF	259,165		88-577	09/03/64
Selway-Bitterroot	Nez Perce NF	559,531	168	88-577	09/03/64
Total Forest Service acres in Idaho:		**3,961,501**	**8,256**		
Geographic State: Illinois					
Bald Knob	Shawnee NF	5,863	55	101-633	11/28/90
Bay Creek	Shawnee NF	2,866		101-633	11/28/90
Burden Falls	Shawnee NF	3,671	52	101-633	11/28/90
Clear Springs	Shawnee NF	4,730		101-633	11/28/90
Garden of the Gods	Shawnee NF	3,268	25	101-633	11/28/90
Lusk Creek	Shawnee NF	4,466	330	101-633	11/28/90
Panther Den	Shawnee NF	685	255	101-633	11/28/90
Total Forest Service acres in Illinois:		**25,549**	**717**		
Geographic State: Indiana					
Charles C. Deam	Hoosier NF	12,935	18	97-384	12/22/82
Total Forest Service acres in Indiana:		**12,935**	**18**		
Geographic State: Kentucky					
Beaver Creek	Daniel Boone NF	4,753	38	93-622	01/03/75
Clifty	Daniel Boone NF	11,662	984	99-197	12/23/85
Total Forest Service acres in Kentucky:		**16,415**	**1,022**		
Geographic State: Louisiana					
Kisatchie Hills	Kisatchie NF	8,679		96-560	12/22/80
Total Forest Service acres in Louisiana:		**8,679**	**0**		
Geographic State: Maine					
Caribou-Speckled Mountain	White Mountain NF	12,000		101-401	09/28/90
Total Forest Service acres in Maine:		**12,000**	**0**		
Geographic State: Michigan					
Big Island Lake	Hiawatha NF	5,856	152	100-184	12/08/87
Delirium	Hiawatha NF	11,870	130	100-184	12/08/87
Horseshoe Bay	Hiawatha NF	3,790	159	100-184	12/08/87
Mackinac	Hiawatha NF	12,230	158	100-184	12/08/87
McCormick	Ottawa NF	16,850		100-184	12/08/87
Nordhouse Dunes	Manistee NF	3,450		100-184	12/08/87
Rock River Canyon	Hiawatha NF	4,640	645	100-184	12/08/87
Round Island	Hiawatha NF	378		100-184	12/08/87
Sturgeon River Gorge	Ottawa NF	14,500	350	100-184	12/08/87
Sylvania	Ottawa NF	18,327		100-184	12/08/87
Total Forest Service acres in Michigan:		**91,891**	**1,594**		
Geographic State: Minnesota					
Boundary Waters Canoe Area	Superior NF	803,050	285,125	88-577	09/03/64
Boundary Waters Canoe Area	Superior NF			95-495	10/21/78
Total Forest Service acres in Minnesota:		**803,050**	**285,125**		
Geographic State: Missouri					
Bell Mountain	Mark Twain NF	8,977	50	96-560	12/22/80
Devils Backbone	Mark Twain NF	6,595		96-560	12/22/80
Hercules-Glades	Mark Twain NF	12,314	1	94-557	10/19/76
Irish	Mark Twain NF	16,117	241	98-289	05/21/84
Paddy Creek	Mark Twain NF	7,019	40	97-407	01/03/83
Piney Creek	Mark Twain NF	8,087	55	96-560	12/22/80
Rockpile Mountain	Mark Twain NF	4,089	42	96-560	12/22/80
Total Forest Service acres in Missouri:		**63,198**	**429**		
Geographic State: Mississippi					
Black Creek	Desoto NF	5,052		98-515	10/19/84
Leaf	Desoto NF	994		98-515	10/19/84
Total Forest Service acres in Mississippi:		**6,046**	**0**		
Geographic State: Montana					
Absaroka-Beartooth	Custer NF	345,589	105	95-249	03/27/78
Absaroka-Beartooth	Gallatin NF	574,738	1,033	95-249	03/27/78
Absaroka-Beartooth	Gallatin NF			98-140	10/31/83
Anaconda-Pintlar	Beaverhead NF	72,677		88-577	09/03/64
Anaconda-Pintlar	Bitterroot NF	41,162		88-577	09/03/64
Anaconda-Pintlar	Deerlodge NF	44,175	642	88-577	09/03/64

Appendix B
Designated Wilderness Areas, Forest Service, 1993

Wilderness Area Name	Administrative Unit Name(s)	Size in Acres (Federal only)	Acres Inholding	Public Law Number	Date(s) of Designation
Bob Marshall	Flathead NF	709,356		88-577	09/03/64
Bob Marshall	Lewis & Clark NF	300,000		88-577	09/03/64
Bob Marshall	Lewis & Clark NF			95-546	10/28/78
Cabinet Mountains	Kaniksu NF	44,320		88-577	09/03/64
Cabinet Mountains	Kootenai NF	49,952		88-577	09/03/64
Gates of the Mountains	Helena NF	28,562		88-577	09/03/64
Great Bear	Flathead NF	286,700		95-546	10/28/78
Lee Metcalf	Beaverhead NF	108,350	1,353	98-140	10/31/83
Lee Metcalf	Gallatin NF	140,594		98-140	10/31/83
Mission Mountains	Flathead NF	73,877		93-632	01/03/75
Rattlesnake	Lolo NF	32,844	156	96-476	10/19/80
Scapegoat	Helena NF	80,697		92-395	08/20/72
Scapegoat	Lewis & Clark NF	84,407		92-395	08/20/72
Scapegoat	Lolo NF	74,192	640	92-395	08/20/72
Selway-Bitterroot	Bitterroot NF	241,676		88-577	09/03/64
Selway-Bitterroot	Lolo NF	9,767		88-577	09/03/64
Welcome Creek	Lolo NF	28,135		95-237	02/24/78
Total Forest Service acres in Montana:		3,371,770	3,929		

Geographic State: North Carolina

Wilderness Area Name	Administrative Unit Name(s)	Size in Acres (Federal only)	Acres Inholding	Public Law Number	Date(s) of Designation
Birkhead Mountains	Uwharrie NF	5,025	135	98-324	06/19/84
Catfish Lake South	Croatan NF	8,530		98-324	06/19/84
Ellicott Rock	Nantahala NF	4,022		93-622	01/03/75
Ellicott Rock	Nantahala NF			98-324	06/19/84
Joyce Kilmer-Slickrock	Nantahala NF	13,562		93-622	01/03/75
Joyce Kilmer-Slickrock	Nantahala NF			98-324	06/19/84
Linville Gorge	Pisgah NF	11,786	216	88-577	09/03/64
Linville Gorge	Pisgah NF			98-324	06/19/84
Middle Prong	Pisgah NF	7,460		98-324	06/19/84
Pocosin	Croatan NF	11,709		98-324	06/19/84
Pond Pine	Croatan NF	1,685		98-324	06/19/84
Sheep Ridge	Croatan NF	9,297		98-324	06/19/84
Shining Rock	Pisgah NF	18,483		88-577	09/03/64
Shining Rock	Pisgah NF			98-324	06/19/84
Southern Nantahala	Nantahala NF	11,703	241	98-324	06/19/84
Total Forest Service acres in North Carolina:		103,262	592		

Geographic State: Nebraska

Wilderness Area Name	Administrative Unit Name(s)	Size in Acres (Federal only)	Acres Inholding	Public Law Number	Date(s) of Designation
Soldier Creek	Nebraska NF	7,794		99-504	10/20/86
Total Forest Service acres in Nebraska:		7,794	0		

Geographic State: New Hampshire

Wilderness Area Name	Administrative Unit Name(s)	Size in Acres (Federal only)	Acres Inholding	Public Law Number	Date(s) of Designation
Great Gulf	White Mountain NF	5,552		88-577	09/03/64
Pemigewasset	White Mountain NF	45,000		98-323	06/19/84
Presidential Range-Dry River	White Mountain NF	27,380		93-622	01/03/75
Presidential Range-Dry River	White Mountain NF			98-323	06/19/84
Sandwich Range	White Mountain NF	25,000		98-323	06/19/84
Total Forest Service acres in New Hampshire:		102,932	0		

Geographic State: New Mexico

Wilderness Area Name	Administrative Unit Name(s)	Size in Acres (Federal only)	Acres Inholding	Public Law Number	Date(s) of Designation
Aldo Leopold	Gila NF	202,016		96-550	12/19/80
Apache Kid	Cibola NF	44,626		96-550	12/19/80
Blue Range	Apache NF	28,104		96-550	12/19/80
Blue Range	Gila NF	1,200		96-550	12/19/80
Capitan Mountains	Lincoln NF	34,658	1,309	96-550	12/19/80
Chama River Canyon	Carson NF	2,900		95-237	02/24/78
Chama River Canyon	Santa Fe NF	47,400		95-237	02/24/78
Cruces Basin	Carson NF	18,000		96-550	12/19/80
Dome	Santa Fe NF	5,200		96-550	12/19/80
Gila	Gila NF	557,873	192	88-577	09/03/64
Gila	Gila NF			96-550	12/19/80
Latir Peak	Carson NF	20,000		96-550	12/19/80
Manzano Mountain	Cibola NF	36,875	320	95-237	02/24/78
Pecos	Carson NF	24,736		88-577	09/03/64
Pecos	Santa Fe NF	198,597		88-577	09/03/64
Pecos	Santa Fe NF			96-550	12/19/80
San Pedro Parks	Santa Fe NF	41,132		88-577	09/03/64
Sandia Mountain	Cibola NF	37,877	480	95-237	02/24/78
Sandia Mountain	Cibola NF			96-248	05/23/80
Sandia Mountain	Cibola NF			98-603	10/30/84
Wheeler Peak	Carson NF	19,661	2	88-577	09/03/64
Wheeler Peak	Carson NF			96-550	12/19/80
White Mountain	Lincoln NF	48,208	677	88-577	09/03/64
White Mountain	Lincoln NF			96-550	12/19/80
Withington	Cibola NF	19,000		96-550	12/19/80
Total Forest Service acres in New Mexico:		1,388,063	2,980		

Geographic State: Nevada

Wilderness Area Name	Administrative Unit Name(s)	Size in Acres (Federal only)	Acres Inholding	Public Law Number	Date(s) of Designation
Alta Toquima	Toiyabe NF	38,000		101-195	12/05/89
Arc Dome	Toiyabe NF	115,000		101-195	12/05/89
Boundary Peak	Inyo NF	10,000		101-195	12/05/89
Currant Mountain	Humboldt NF	36,000		101-195	12/05/89
East Humboldts	Humboldt NF	36,900		101-195	12/05/89
Grant Range	Humboldt NF	50,000		101-195	12/05/89
Jarbidge	Humboldt NF	113,167	160	88-577	09/03/64
Jarbidge	Humboldt NF			101-195	12/05/89
Mount Charleston	Toiyabe NF	43,000		101-195	12/05/89
Mount Moriah	Humboldt NF	70,000		101-195	12/05/89
Mount Rose	Toiyabe NF	28,000		101-195	12/05/89
Quinn Canyon	Humboldt NF	27,000		101-195	12/05/89

Appendix B
Designated Wilderness Areas, Forest Service, 1993

Wilderness Area Name	Administrative Unit Name(s)	Size in Acres (Federal only)	Acres Inholding	Public Law Number	Date(s) of Designation
Ruby Mountains	Humboldt NF	90,000		101-195	12/05/89
Santa Rosa - Paradise Peak	Humboldt NF	31,000		101-195	12/05/89
Table Mountain	Toiyabe NF	98,000		101-195	12/05/89
Total Forest Service acres in Nevada:		786,067	160		
Geographic State: Oklahoma					
Black Fork Mountain	Ouachita NF	4,549	600	100-499	10/18/88
Upper Kiamichi River	Ouachita NF	9,602	1,217	100-499	10/18/88
Total Forest Service acres in Oklahoma:		14,151	1,817		
Geographic State: Oregon					
Badger Creek	Mt. Hood NF	24,000		98-328	06/26/84
Black Canyon	Ochoco NF	13,400		98-328	06/26/84
Boulder Creek	Umpqua NF	19,100		98-328	06/26/84
Bridge Creek	Ochoco NF	5,400		98-328	06/26/84
Bull of the Woods	Mt. Hood NF	27,427		98-328	06/26/84
Bull of the Woods	Willamette NF	7,473		98-328	06/26/84
Columbia	Mt. Hood NF	39,000		98-328	06/26/84
Cummins Creek	Siuslaw NF	9,173		98-328	06/26/84
Diamond Peak	Deschutes NF	34,413		88-577	09/03/64
Diamond Peak	Deschutes NF			98-328	06/26/84
Diamond Peak	Willamette NF	19,772		88-577	09/03/64
Diamond Peak	Willamette NF			98-328	06/26/84
Drift Creek	Siuslaw NF	5,798		98-328	06/26/84
Eagle Cap	Wallowa NF	212,699	1,149	88-577	09/03/64
Eagle Cap	Wallowa NF			92-521	10/21/72
Eagle Cap	Wallowa NF			98-328	06/26/84
Eagle Cap	Whitman NF	145,762	665	88-577	09/03/64
Eagle Cap	Whitman NF			92-521	10/21/72
Eagle Cap	Whitman NF			98-328	06/26/84
Gearhart Mountain	Fremont NF	22,809		88-577	09/03/64
Gearhart Mountain	Fremont NF			98-328	06/26/84
Grassy Knob	Siskiyou NF	17,200		98-328	06/26/84
Hells Canyon	Wallowa NF	118,247		94-199	12/31/75
Hells Canyon	Wallowa NF			98-328	06/26/84
Hells Canyon	Whitman NF	11,848	1,038	94-199	12/31/75
Hells Canyon	Whitman NF			98-328	06/26/84
Kalmiopsis	Siskiyou NF	179,655	45	88-577	09/03/64
Kalmiopsis	Siskiyou NF			95-237	02/24/78
Menagerie	Willamette NF	4,800		98-328	06/26/84
Middle Santiam	Willamette NF	7,500		98-328	06/26/84
Mill Creek	Ochoco NF	17,400		98-328	06/26/84
Monument Rock	Malheur NF	12,620		98-328	06/26/84
Monument Rock	Whitman NF	7,030		98-328	06/26/84
Mount Hood	Mt. Hood NF	46,520	640	88-577	09/03/64
Mount Hood	Mt. Hood NF			95-237	02/24/78
Mount Jefferson	Deschutes NF	32,734		90-548	10/02/68
Mount Jefferson	Mt. Hood NF	5,021		90-548	10/02/68
Mount Jefferson	Willamette NF	69,253		90-548	10/02/68
Mount Jefferson	Willamette NF			98-328	06/26/84
Mount Thielsen	Deschutes NF	7,107		98-328	06/26/84
Mount Thielsen	Umpqua NF	21,593		98-328	06/26/84
Mount Thielsen	Winema NF	25,567		98-328	06/26/84
Mount Washington	Deschutes NF	14,116		88-577	09/03/64
Mount Washington	Deschutes NF			98-328	06/26/84
Mount Washington	Willamette NF	38,622		88-577	09/03/64
Mount Washington	Willamette NF			98-328	06/26/84
Mountain Lakes	Winema NF	23,071		88-577	09/03/64
North Fork John Day	Umatilla NF	107,058		98-328	06/26/84
North Fork John Day	Whitman NF	14,294		98-328	06/26/84
North Fork Umatilla	Umatilla NF	20,435		98-328	06/26/84
Red Buttes	Rogue River NF	350		98-425	09/28/84
Red Buttes	Siskiyou NF	3,400		98-328	06/26/84
Rock Creek	Siuslaw NF	7,486		98-328	06/26/84
Rogue-Umpqua Divide	Rogue River NF	6,850		98-328	06/26/84
Rogue-Umpqua Divide	Umpqua NF	26,350		98-328	06/26/84
Salmon-Huckleberry	Mt. Hood NF	44,560	40	98-328	06/26/84
Sky Lakes	Rogue River NF	75,695		98-328	06/26/84
Sky Lakes	Winema NF	40,605		98-328	06/26/84
Strawberry Mountain	Malheur NF	68,700	650	88-577	09/03/64
Strawberry Mountain	Malheur NF			98-328	06/26/84
Three Sisters	Deschutes NF	94,370		88-577	09/03/64
Three Sisters	Deschutes NF			98-328	06/26/84
Three Sisters	Willamette NF	192,338		88-577	09/03/64
Three Sisters	Willamette NF			95-237	02/24/78
Three Sisters	Willamette NF			98-328	06/26/84
Waldo Lake	Willamette NF	39,200		98-328	06/26/84
Wenaha-Tucannon	Umatilla NF	66,375	42	95-237	02/24/78
Wild Rogue	Siskiyou NF	34,629	11,342	95-237	02/24/78
Total Forest Service acres in Oregon:		2,088,825	15,611		
Geographic State: Pennsylvania					
Allegheny Islands	Allegheny NF	368		98-585	10/30/84
Hickory Creek	Allegheny NF	8,570		98-585	10/30/84
Total Forest Service acres in Pennsylvania:		8,938	0		
Geographic State: South Carolina					
Ellicott Rock	Sumter NF	2,809		93-622	01/03/75
Hell Hole Bay	Francis Marion NF	2,180		96-560	12/22/80
Little Wambaw Swamp	Francis Marion NF	5,154		96-560	12/22/80
Wambaw Creek	Francis Marion NF	1,937		96-560	12/22/80
Wambaw Swamp	Francis Marion NF	4,767		96-560	12/22/80

Appendix B
Designated Wilderness Areas, Forest Service, 1993

Wilderness Area Name	Administrative Unit Name(s)	Size in Acres (Federal only)	Acres Inholding	Public Law Number	Date(s) of Designation
Total Forest Service acres in South Carolina:		16,847	0		
Geographic State: South Dakota					
Black Elk	Black Hills NF	9,826		96-560	12/22/80
Total Forest Service acres in South Dakota:		9,826	0		
Geographic State: Tennessee					
Bald River Gorge	Cherokee NF	3,721		98-578	10/30/84
Big Frog	Cherokee NF	7,986		98-578	10/30/84
Big Frog	Cherokee NF			99-490	10/16/86
Big Laurel Branch	Cherokee NF	6,251		99-490	10/16/86
Citico Creek	Cherokee NF	16,226		98-578	10/30/84
Cohutta	Cherokee NF	1,795		93-622	01/03/75
Gee Creek	Cherokee NF	2,493		93-622	01/03/75
Joyce Kilmer-Slickrock	Cherokee NF	3,832		93-622	01/03/75
Little Frog Mountain	Cherokee NF	4,684		99-490	10/16/86
Pond Mountain	Cherokee NF	6,626	39	99-490	10/16/86
Sampson Mountain	Cherokee NF	7,991	1	99-490	10/16/86
Unaka Mountain	Cherokee NF	4,700		99-490	10/16/86
Total Forest Service acres in Tennessee:		66,305	40		
Geographic State: Texas					
Big Slough	Davy Crockett NF	3,455		98-574	10/30/84
Indian Mounds	Sabine NF	9,358	1,559	98-574	10/30/84
Indian Mounds	Sabine NF			99-584	10/29/86
Little Lake Creek	Sam Houston NF	3,855		98-574	10/30/84
Little Lake Creek	Sam Houston NF			99-584	10/29/86
Turkey Hill	Angelina NF	5,473		98-574	10/30/84
Turkey Hill	Angelina NF			99-584	10/29/86
Upland Island	Angelina NF	12,571	79	98-574	10/30/84
Upland Island	Angelina NF			99-584	10/29/86
Total Forest Service acres in Texas:		34,712	1,638		
Geographic State: Utah					
Ashdown Gorge	Dixie NF	7,000		98-428	09/18/84
Box-Death Hollow	Dixie NF	25,751	63	98-428	09/18/84
Dark Canyon	Manti-La Sal NF	45,000		98-428	09/18/84
Deseret Peak	Wasatch NF	25,500		98-428	09/18/84
High Uintas	Ashley NF	276,175		98-428	09/18/84
High Uintas	Wasatch NF	180,530		98-428	09/18/84
Lone Peak	Uinta NF	21,166		95-237	02/24/78
Lone Peak	Wasatch NF	8,922		95-237	02/24/78
Mount Naomi	Cache NF	44,350		98-428	09/18/84
Mount Nebo	Uinta NF	28,000		98-428	09/18/84
Mount Olympus	Wasatch NF	16,000		98-428	09/18/84
Mount Timpanogos	Uinta NF	10,750		98-428	09/18/84
Pine Valley Mountain	Dixie NF	50,000		98-428	09/18/84
Twin Peaks	Wasatch NF	11,334	129	98-428	09/18/84
Wellsville Mountain	Cache NF	23,850		98-428	09/18/84
Total Forest Service acres in Utah:		774,328	192		
Geographic State: Virginia					
Barbours Creek	George Washington NF	5		100-326	06/07/88
Barbours Creek	Jefferson NF	5,695		100-326	06/07/88
Beartown	Jefferson NF	5,609		98-586	10/30/84
James River Face	Jefferson NF	8,886		93-622	01/03/75
James River Face	Jefferson NF			98-586	10/30/84
Kimberling Creek	Jefferson NF	5,542		98-586	10/30/84
Lewis Fork	Jefferson NF	5,618		98-586	10/30/84
Little Dry Run	Jefferson NF	2,858		98-586	10/30/84
Little Wilson Creek	Jefferson NF	3,613		98-586	10/30/84
Mountain Lake	Jefferson NF	8,187	205	98-586	10/30/84
Peters Mountain	Jefferson NF	3,328		98-586	10/30/84
Ramseys Draft	George Washington NF	6,518		98-586	10/30/84
Rich Hole	George Washington NF	6,450		100-326	06/07/88
Rough Mountain	George Washington NF	9,300		100-326	06/07/88
Saint Mary's	George Washington NF	9,835		98-586	10/30/84
Shawvers Run	George Washington NF	95		100-326	06/07/88
Shawvers Run	Jefferson NF	3,570		100-326	06/07/88
Thunder Ridge	Jefferson NF	2,344		98-586	10/30/84
Total Forest Service acres in Virginia:		87,453	205		
Geographic State: Vermont					
Big Branch	Green Mountain NF	6,720		98-322	06/19/84
Breadloaf	Green Mountain NF	21,480		98-322	06/19/84
Bristol Cliffs	Green Mountain NF	3,738		93-622	01/03/75
Bristol Cliffs	Green Mountain NF			94-268	04/16/76
George D. Aiken	Green Mountain NF	5,060		98-322	06/19/84
Lye Brook	Green Mountain NF	15,503	177	93-622	01/03/75
Lye Brook	Green Mountain NF			98-322	06/19/84
Peru Peak	Green Mountain NF	6,920		98-322	06/19/84
Total Forest Service acres in Vermont:		59,421	177		
Geographic State: Washington					
Alpine Lakes	Snoqualmie NF	117,776	123	94-357	07/12/76
Alpine Lakes	Wenatchee NF	244,845	1,485	94-357	07/12/76

Appendix B
Designated Wilderness Areas, Forest Service, 1993

Wilderness Area Name	Administrative Unit Name(s)	Size in Acres (Federal only)	Acres Inholding	Public Law Number	Date(s) of Designation
Boulder River	Mt. Baker NF	48,674		98-339	07/03/84
Buckhorn	Olympic NF	44,258	216	98-339	07/03/84
Buckhorn	Olympic NF			99-635	11/07/86
Clearwater	Snoqualmie NF	14,374		98-339	07/03/84
Colonel Bob	Olympic NF	11,961		98-339	07/03/84
Glacier Peak	Mt. Baker NF	283,252	252	88-577	09/03/64
Glacier Peak	Mt. Baker NF			90-544	10/02/68
Glacier Peak	Mt. Baker NF			98-339	07/03/84
Glacier Peak	Wenatchee NF	289,086	148	88-577	09/03/64
Glacier Peak	Wenatchee NF			90-544	10/02/68
Glacier Peak	Wenatchee NF			98-339	07/03/84
Glacier View	Gifford Pinchot NF	3,123		98-339	07/03/84
Goat Rocks	Gifford Pinchot NF	71,203		88-577	09/03/64
Goat Rocks	Gifford Pinchot NF			98-339	07/03/84
Goat Rocks	Snoqualmie NF	37,076	160	88-577	09/03/64
Goat Rocks	Snoqualmie NF			98-339	07/03/84
Henry M. Jackson	Mt. Baker NF	27,985	460	98-339	07/03/84
Henry M. Jackson	Snoqualmie NF	47,446	460	98-339	07/03/84
Henry M. Jackson	Wenatchee NF	27,242		98-339	07/03/84
Indian Heaven	Gifford Pinchot NF	20,960		98-339	07/03/84
Lake Chelan-Sawtooth	Okanogan NF	95,021	87	98-339	07/03/84
Lake Chelan-Sawtooth	Wenatchee NF	56,414	42	98-339	07/03/84
Mount Adams	Gifford Pinchot NF	46,626	10,055	88-577	09/03/64
Mount Adams	Gifford Pinchot NF			98-339	07/03/84
Mount Baker	Mt. Baker NF	117,528	320	98-339	07/03/84
Mount Skokomish	Olympic NF	13,015		98-339	07/03/84
Mount Skokomish	Olympic NF			99-635	11/07/86
Noisy-Diobsud	Mt. Baker NF	14,133		98-339	07/03/84
Norse Peak	Snoqualmie NF	52,180	21	98-339	07/03/84
Pasayten	Mt. Baker NF	107,039		90-544	10/02/68
Pasayten	Okanogan NF	422,992		90-544	10/02/68
Pasayten	Okanogan NF			98-339	07/03/84
Salmo-Priest	Colville NF	29,386		98-339	07/03/84
Salmo-Priest	Kaniksu NF	11,949		98-339	07/03/84
Tatoosh	Gifford Pinchot NF	15,750		98-339	07/03/84
The Brothers	Olympic NF	16,682		98-339	07/03/84
The Brothers	Olympic NF			99-635	11/07/86
Trapper Creek	Gifford Pinchot NF	5,970		98-339	07/03/84
Wenaha-Tucannon	Umatilla NF	111,048		95-237	02/24/78
William O. Douglas	Gifford Pinchot NF	15,469		98-339	07/03/84
William O. Douglas	Snoqualmie NF	152,688		98-339	07/03/84
Wonder Mountain	Olympic NF	2,349		98-339	07/03/84
Total Forest Service acres in Washington:		**2,575,500**	**13,829**		

Geographic State: Wisconsin

Wilderness Area Name	Administrative Unit Name(s)	Size in Acres (Federal only)	Acres Inholding	Public Law Number	Date(s) of Designation
Blackjack Springs	Nicolet NF	5,886		95-494	10/21/78
Headwaters	Nicolet NF	18,188	1,916	98-321	06/19/84
Porcupine Lake	Chequamegon NF	4,292	154	98-321	06/19/84
Rainbow Lake	Chequamegon NF	6,583		93-622	01/03/75
Whisker Lake	Nicolet NF	7,428	83	95-494	10/21/78
Total Forest Service Acres in Wisconsin		**44,447**	**2,153**		

Geographic State: West Virginia

Wilderness Area Name	Administrative Unit Name	Size in Acres	Acres Inholding	Public Law Number	Date(s) of Designation
Cranberry	Monongahela NF	35,864		97-466	01/13/83
Cranberry	Monongahela NF			101-512	11/05/90
Dolly Sods	Monongahela NF	10,215		93-622	01/03/75
Laurel Fork North	Monongahela NF	6,055		97-466	01/13/83
Laurel Fork South	Monongahela NF	5,997		97-466	01/13/83
Mountain Lake	Jefferson NF	2,721		100-326	06/07/88
Otter Creek	Monongahela NF	20,000		93-622	01/03/75
Total Forest Service Acres in West Virginia		**80,852**	**0**		

Geographic State: Wyoming

Wilderness Area Name	Administrative Unit Name	Size in Acres	Acres Inholding	Public Law Number	Date(s) of Designation
Absaroka-Beartooth	Shoshone NF	23,283		98-550	10/30/84
Bridger	Bridger NF	428,087		88-577	09/03/64
Bridger	Bridger NF	189,039		98-550	10/30/84
Cloud Peak	Bighorn NF	10,124		98-550	10/30/84
Encampment River	Medicine Bow NF	198,525		94-557	10/19/76
Fitzpatrick	Shoshone NF			94-567	10/20/76
Fitzpatrick	Shoshone NF			98-550	10/30/84
Fitzpatrick	Shoshone NF			98-550	10/30/84
Gros Ventre	Teton NF	287,000		98-550	10/30/84
Huston Park	Medicine Bow NF	30,726	138	98-550	10/30/84
Jedediah Smith	Targhee NF	123,451		88-577	09/03/64
North Absaroka	Shoshone NF	350,488		98-550	10/30/84
Platte River	Medicine Bow NF	22,749		98-550	10/30/84
Popo Agie	Shoshone NF	101,870		95-237	02/24/78
Savage Run	Medicine Bow NF	14,930	3	88-577	09/03/64
Teton	Teton NF	585,238		98-550	10/30/84
Teton	Teton NF			88-577	09/03/64
Washakie	Shoshone NF	704,822	548	98-550	10/30/84
Washakie	Shoshone NF			98-550	10/30/84
Washakie	Shoshone NF			98-550	10/30/84
Winegar Hole	Targhee NF	10,715			
Total Forest Service Acres in Wyoming		**3,081,047**	**689**		

Total Forest Service acres in the United States 34,017,024   451,049

Note: Multiple listings are included for those areas designated or affected by more than one Public Law; managed by more than one agency; located in more than one administrative unit.

REFERENCE NOTES

Chapter I

[1] Joel Gottlieb, "The Preservation of Wilderness Values: The Politics and Administration of Conservation Policy," Ph.D. dissertation, University of California at Riverside, 1972, pp. 121-32.

[2] Stephen Fox's John Muir and His Legacy: The American Conservation Movement (Boston: Little, Brown & Company, 1981) offers a good general history of the conservation movement. Also useful for this study has been Douglas Scott's "Origins and Development of the Wilderness Bill, 1930-1956." Mr. Scott was kind enough to send me a copy of this rough draft of his University of Michigan master's thesis.

[3] Elliot Barker to Senator Clinton B. Anderson, August 12, 1959, box 619, Clinton B. Anderson Papers, Manuscript Reading Room, Library of Congress.

[4] Susan L. Flader, Thinking Like a Mountain: Aldo Leopold and the Evolution of an Ecological Attitude toward Deer, Wolves, and Forests (Columbia: University of Missouri Press, 1974), pp. 15-17.

[5] James P. Gilligan, "The Development of Policy and Administration of Forest Service Primitive and Wilderness Areas in the Western United States," Ph.D. dissertation. University of Michigan, 1953, p. 77.

[6] Ibid., p. 83; Donald Nicholas Baldwin, The Quiet Revolution: Grass Roots of Today's Wilderness Preservation Movement (Boulder: Pruett Publishing Company, 1972), pp. 153-65.

[7] Baldwin, Quiet Revolution, p. 34.

[8] Carhart quoted in Walter Gallacher, The White River National Forest, 1891-1981 (Glenwood Springs, Colorado: White River National Forest, 1981), p. 45.

[9] U.S. Forest Service, Report of the Forester (Washington, D.C.: GPO, 1926), p. 36; ibid. (1927), p. 13.

[10] Regulation L-20, October 1930.

[11] Gilligan, "Development of Policy." pp. 191-97.

[12] Gottlieb, "Preservation of Wilderness Values," pp. 141-53.

[13] Richard E. McArdle, "Wilderness Politics: Legislation and Forest Servicy Policy," Journal of Forest History 19 (October 1978):166-79.

[14] Gilligan, "Development of Policy," pp. 118-98.

[15] Ibid, p. 157.

[16] Scott, "Origins and Development," chapter 4, pp. 10-21.

[17] Ibid, chapter 3, pp. 28-35.

[18] Gilligan, "Development of Policy," p. 197.

[19] Ibid.

[20] Jack M. Hession, "The Legislative History of the Wilderness Act," M.A. thesis, San Diego State University, 1967, p. 27

[21] Scott, "Origins and Development," chapter 2, pp. 16-22; Michael McCloskey, "The Wilderness Act of 1964: Its Background and Meaning," Oregon Law Review 45 (no 4, 1966): 294.

[22] Gilligan, "Development of Policy," pp. 218-38, 263-54.

[23] Hession, "Legislative History of the Wilderness Act," p. 31

[24] Living Wilderness 11 (Winter 1947-1948):1.

[25] Albert Dixon, "The Conservation of Wilderness: A Study in Politics," Ph.D. dissertation, University of California at Berkeley, 1968, p. 61.

[26] U.S. Department of Agriculture, The Principal Laws Relating to Forest Service Activities, Agricultural Handbook no. 453 (Washington D.C.: GPO, 1978), p. 201.

[27] Living Wilderness 11 (December 1964):28.

[28] Zahniser to Clinton B. Anderson, August 6, 1959, box 619, Anderson Papers.

[29] Living Wilderness 12 (Winter 1947-1948): 1; Fox, John Muir, pp. 269-71.

[30] Fox, John Muir, pp. 267, 269; Living Wilderness 12 (Autumn 1947): 7-16.

[31] Living Wilderness 12 (Winter 1947-1948): 1; Ibid. (Autumn 1947): 12.

[32] Ibid. (Winter 1947-1948):4.

[33] Fox, John Muir, p. 287.

[34] Ibid.; Living Wilderness 12 (Winter 1947-1948): 4.

[35] For details on the drafting of the wilderness bill see Scott, "Origins and Development," chapters 5-7.

[36] Sundquist, Politics and Policy: The Eisenhower, Kennedy, and Johnson Years (Washington, D.C.: Brookings Institution, 1958), p. 337.

[37] McCloskey, "Wilderness Act," p. 298.

[38] Letter dated March 15, 1963, Senate Interior and Insular Affairs Committee, S.2 S.4 (cont.), Record Group 46, National Archives.

[39] Harold K. Steen, The U.S. Forest Service: A History (Seattle: University of Washington Press, 1976), pp. 301-07.

[40] Congressional Record, 86 Cong., 2d sess., July 2, 1960, p. 15566; ibid., 87 Cong., 1st sess., September 5, 1961, p. 18090.

[41] Living Wilderness 21 (Winter-Spring 1956-1957): 34; U.S. Senate

[42] Living Wilderness 21 (Winter-Spring 1956-1957): 34; U.S. Senate, Establishing a National Wilderness Preservation System, 88 Cong., 1st sess., SR 109, 1963, p. 10; U.S. Senate, Hearings Before the Committee on interior and Insular Affairs, 85 Cong., lst sess., June 19, 29, 1957 (hereinafter cited as Hearings), p. 181-82.

[43] U.S. Senate, Committee on Interior and Insular Affairs, National Wilderness Preservation Act, Hearings, 85 Cong. 2d sess., 1959, p. 310.

[44] Hearings, p. 394; Living Wilderness 22 (Autumn 1957): 31.

[45] Congressional Record, 87 Cong., lst sess., September 5, 1961, p. 18089.

[46] Congressional Record, 88 Cong., 2d sess., August 20, 1964, p. 20601.

[47] Hession, "Legislative History of the Wilderness Act"; Roderick Nash, Wilderness and the American Mind, 3rd edition (New Haven: Yale University press, 1982); and Dixon, "Conservtion of Wilderness."

[48] U.S. House of Representatives, Hearings Before the Subcommittee on Public Lands, 88 Cong., 2d sess., April 27-30, May 1, 1964, p. 1227.

Chapter II

[1] "Executive Director's Report to Council, 1966-67", Denver Conservation Library (DCL), Wilderness Society Records (WSR), Box 2.

[2] Telephone interview with Richard Costley, 4/27/83, Forest Service History Section (FSHS).

[3] Interview with Ernest Dickerman, Buffalo Gap, VA, 8/18/83, FSHS.

[4] FSHS, photocopy of Senate Hearings on S. 316, 1973, p. 31-35.

[5] Interview with John Hall, Washington, DC, 7/25/83, FSHS.

[6] Brock Evans, "The Wilderness Idea as a Moving Force in American Cultural and Political

History," Congressional Record 4/27/81.
[7] Interview with Brock Evans, Washington, DC, 7/15/83, FSHS.
[8] William Worf, "Two Faces of Wilderness - A Time for Choice," Idaho Law Review, Vol. 16, No. 3 (Summer 1980), p. 435.
[9] Richard Costley, letter to J.W. Deinema, 4/19/71, FSHS.
[10] "National Forest Wilderness: A Policy Review", Forest Service, 1972, page 5.
[11] Richard J. Costley, "An Enduring Resource," American Forests, October 1971.
[12] Remarks of Art Greeley to Regional Foresters," Ogden, UT 12.10/64, Forest Service, Washington Office, Division of Recreation Records.
[13] Speech by Rupert Cutler, 8/20/68 DCL, WSR, Box 2, Evans Interview, cited above.
[14] Costley letter to J.W. Deinema, cited above.
[15] Ibid.
[16] Ibid.
[17] Bill Worf, letter to the author 5/15/84, FSHS.
[18] DCL, WSR, Box 2, undated draft letter.
[19] Michael McCloskey, "The Wilderness Act of 1964: Its Background and Meaning," Oregon Law Review, Vol. 45, No. 4 (June 1966), p. 295.
[20] Costley letter to J.W. Deinema, cited above.
[21] Interview with Harry Crandell, Washington, DC, 5/20/83, FSHS.
[22] "National Forest Wilderness: A Policy Review," cited above, p. 48.
[23] Dickerman interview, cited above.
[24] John C. Hendee, George H. Stankey, and Robert C. Lucas, Wilderness Management (Washington, DC: USDA Forest Service, 1978), pp. 249-251.

Chapter III

[1] Interview with Clifton Merritt, Denver, CO, 6/23/83, Forest Service History Section (FSHS).
[2] Ferdinand Silcox, letter to Warren Eaton, 11/8/35, Denver Conservation Library, (DCL), Wilderness Society Records (WSL), San Rafael, Scapegoat and other Wilderness Areas Box.
[3] Telephone interview with Richard Costley, 4/27/83, FSHS.
[4] Forest Service, San Francisco, Division of Recreation Records, 12/3/65.
[5] Ibid.
[6] Ibid.
[7] "Release to Cooperators, 4/4/67," DCL, WSR, Box 2.
[8] Bill Worf, letter to the author, 5/15/84, FSHS.
[9] Interview with Rupert Cutler, New York, NY, 7/21/83, FSHS.
[10] Interview with Brock Evans, Washington, DC, 7/15/83, FSHS.
[11] Quoted in Ronald Gibson Strickland's "Ten Years of Congressional Review Under the Wilderness Act of 1964: Wilderness Classification Through Affirmative Action", Ph.D. dissertation, Georgetown University, 1976, p. 56.
[12] Congressional Record, 3/5/68, p 5238.
[13] Congressional Record, 3/5/68, p. 5242.
[14] Cutler interview, cited above.
[15] Ibid.
[16] Ibid.
[17] Interview with John McGuire, Washington, DC, 6/16/83, FSHS.
[18] Strickland, "Ten Years of Congressional Review," cited above, p. 77.
[19] Doug Scott letter to Senator Floyd Haskell, 2/27/73, DCL, WSR, Eastern Wilderness Box.
[20] Report on H.R. 3454 Report No. 95-490, Senate Committee on Energy and Natural Resources, 1977, p. 7.
[21] Interview with Dick Joy, Washington, DC, 6/16/83, FSHS.
[22] Chief Edward Cliff, "San Gabriel, Washakie, and Mount Jefferson Wilderness Areas," Hearings Before the Subcommittee on Public Lands, U.S. Senate, 90th Cong. 2nd Session, 1968, p. 17.
[23] Chief Edward Cliff, "San Gabriel, Washakie, and Mount Jefferson Wilderness Areas," Hearings Before the Subcommittee on Public Lands, U.S. Senate, 90th Cong. 2nd Session, 1968, p. 17.
[24] "Biological Evaluation R2-81-6," Forest Service, State and Private Forestry, Denver, Colorado, 1981.
[25] DCL, WSR, Box 45, Wyoming State Journal, 2/12/68.
[26] DCL, WSR, Box 45, "Sierra Club Study of the Washakie," 8/66.
[27] Thomas Bell, letter to Joe Green, 10/8/69, DCL, WSR, Box 45.
[28] DCL, WSR, Box 45, Wyoming State Journal, 11/28/68.
[29] William Worf, letter to William Isaacs, 10/13/67, DCL, WSR, Box 45.
[30] Thomas Bell, letter to Joe Green, 10/8/69, DCL, WSR, Box 45.
[31] Interview with Tim Mahoney, Washington, DC, 6/15/83, FSHS.
[32] Gale McGee, letter to Keith Becker, 9/15/71. DCL, WSR, Box 45.
[33] Clifton Merritt, letter to Orrin Bonney, 4/15/71, DCL, WSR, Box 45.
[34] Thomas Bell, letter to Clifford Hansen, 9/22/71, DCL, WSR, Box 45.
[35] Clifford Hansen, letter to Thomas Bell, 10/25/71, DCL, WSR, Box 45.
[36] Interview with Andy Wiessner, Washington, DC, 5/20/83, FSHS.
[37] Malcolm Rupert Cutler, "A Study of Litigation Related to Management of Forest Service Administered Lands and its Effect on Policy Decisions", Ph.D. dissertation, Michigan State University, 1972, pp. 1-12.
[38] Ibid.
[39] Ibid.
[40] Ibid.
[41] Ibid.
[42] Merritt interview, cited above.
[43] Clifton Merritt, letter to Steward Brandborg, 5/2/69, DCL, WSR, Box 42.
[44] Frank J. Barry, Letter to Steward Brandborg, 5/20/69, DCL, WSR, Box 42.
[45] Merritt Interview, cited above.
[46] Ibid.
[47] Telephone interview with Tony Ruckel, 7/27/83, FSHS.
[48] Cutler, "Litigation," cited above, p. 103.
[49] Records of John McGuire, Parker Case, FSHS.

Chapter IV

[1] R.W. Behan, "The Lincoln Back Country Controversy", unpublished ms., Forest Service History Section (FSHS), 1965.
[2] Records of Robert Morgan, memo from Western Montana Conservation Coaliton, 1/27/65.
[3] Interview with Clifton Merritt, Denver, CO, 6/23/83, FSHS.
[4] Behan, "Lincoln Back Country", cited above.
[5] Ibid.

[6] Ibid.
[7] Ibid.
[8] Chief Edward Cliff, letter to Senator Mike Mansfield, 6/28/62, Records of Robert Morgan.
[9] Behan, "Lincoln Back Country", cited above.
[10] Robert Morgan, letter to Regional Office, 2/24/64, Records of Robert Morgan.
[11] W. M. Dreskill, letter to Boyd Rasmussen, 8/20/62, Records of Robert Morgan.
[12] Ibid.
[13] Denver Conservation Library (DCL), Wilderness Society Records (WSR), San Rafael, Scapegoat, and other Wilderness Areas Box, Testimony of Thomas Edwards, 4/16/69.
[14] Behan, "Lincoln Back Country", cited above.
[15] Cecil Garland, reminiscences, FSHS.
[16] Behan, "Lincoln Back Country", cited above.
[17] Records of Robert Morgan, Great Falls Tribune, 10/16/63.
[18] Robert Morgan, letter to the Regional Office, 1/17/64, Records of Robert Morgan.
[19] Merritt interview, cited above.
[20] Ibid.
[21] Robert Morgan, letter to Regional Office, 2/24/64, Records of Robert Morgan.
[22] Robert Morgan, letter to Regional Office, 8/11/64, Records of Robert Morgan.
[23] Merritt interview, cited above.
[24] Ibid.
[25] Ibid.
[26] Richard Johnson, letter to Wayne Aspinall, 11/20/69, DCL, WSR, San Rafael, Scapegoat, and other Wilderness Areas Box.
[27] Merritt interview.
[28] Neil Rahm, letter to Ed Cliff, 2/23/72, Records of Robert Morgan.
[29] Speech by Neil M. Rahm, 3/13/69, Records of Robert Morgan.
[30] Neil Rahm, letter to Ed Cliff, 2/13/71, Records of Robert Morgan.
[31] Cecil Garland reminiscences, cited above.
[32] Ronald Gibson Strickland, "Ten Years of Congressional Review under the Wilderness Act of 1964: Wilderness Classification through Affirmative Action", Ph.D. dissertation, Georgetown University, 1976, p. 61.

Chapter V

[1] Interview with Dick Joy, Washington, DC, 6/16/83, Forest Service History Section (FSHS).
[2] Interview with Edward Cliff, Washington, DC, 9/17/83, FSHS.
[3] Interview with John McGuire, Washington, DC, 6/16/83, FSHS.
[4] Interview with Brock Evans, Washington, DC, 7/15/83, FSHS.
[5] Cliff and McGuire interviews, cited above.
[6] Ronald Gibson Strickland, "Ten Years of Congressional Review under the Wilderness Act of 1964: Wilderness Classification Through Affirmative Action," Ph.D. Dissertation, Georgetown University 1976, p. 61.
[7] Steward Brandborg, letter to James Marshall, 7/10/72, DCL, WSR, Box 39.
[8] "Northwest Conservation Brief," 9/72, Denver Conservation Library (DCL), Wilderness Society Records (WSR), Box 39.
[9] Strickland, "Ten Years of Congressional Review," cited above, pp. 300-201.
[10] Interview with Clifton Merritt, Denver, CO, 6/23/83, FSHS.
[11] Interview with Rupert Cutler, New York, NY, 7/21/83, FSHS.
[12] "West Virginia Organization," DCL, WSR, Box 44.
[13] George Langford, letter to Rupert Cutler, 7/7/69, DCL, WSR, Box 44.
[14] "Eastern Wilderness Flash," 4/24/72, DCL, WSR, Box 5.
[15] Interview with Max Peterson, Washington, DC, 2/6/84, FSHS.
[16] Forest Service, Washington Office, Division of Recreation Records, "Position Paper," 12.9/71.
[17] Forest Service, Washington Office, Division of Recreation Records, 9/24/71.
[18] Interview with Douglas Scott, San Francisco, CA, 9/18/83, FSHS.
[19] DCL, WSR, Eastern Wilderness Box, Forest Service.
[20] Note from Harry Crandell to the author, no date, FSHS.
[21] Strickland, "Ten Years of Congressional Review," cited above, p. 186.
[22] McGuire interview, cited above; "Flyer on Eastern Areas Wilderness Bill," 2/73, DCL, WSR, Box 39.
[23] Scott Interview, cited above.
[24] Letter from Ernie Dickerman to the author, 3/24/84, FSHS.
[25] McGuire interview, cited above.
[26] Scott interview, cited above.
[27] Senator Hermann Talmadge, letter to Senator Paul Fannin, 10/2/73, DCL, WSR, Eastern Wilderness Box.
[28] McGuire interview, cited above.,
[29] Interview with Ernest Dickerman, Buffalo Gap, VA, 8/18/83, FSHS.
[30] Scott inteview, cited above.
[31] DCL, WSR, Eastern Wilderness Box, Congressional Record, 6/13/72.
[32] Scott interview, cited above.
[33] Dickerman interview, cited above.
[34] Ernie Dickerman, letter to Hal Scott, 9/19/72, DCL, WSR, Eastern Wilderness Box.
[35] Cutler interview, cited above.
[36] DCL, WSR, Box 6, "Report on meeting with Forest Service," 3/10/72.
[37] Scott interview, cited above.
[38] Helen McGinnis, letter to George Alderson, 1/9/73, DCL, WSR, Box 6.
[39] Scott interview, cited above.
[40] Ibid.
[41] George Alderson, "Eastern Wilderness Crisis," Environmental Quality, Vol. 4, No. 3 (March 1973), p. 72.
[42] Scott interview, cited above.
[43] "Eastern Wilderness Areas Act," 93rd Cong. 1st session, Senate Report No. 93-661, 1973; Senator Henry Jackson, letter to Senator Hermann Talmadge, no date, DCL, WSR, Eastern Wilderness Box.
[44] Dickerman interview, cited above.
[45] Ernie Dickerman, letter to Senator Henry Jackson, 12/17/74, DCL, WSR, Eastern Wilderness Box.

Chapter VI

[1] Interview with Gene Bergofften, Washington, DC, 7/25/83, Forest Service History Section (FSHE).
[2] Interview with Douglas Scott, San Francisco, CA, 9/18/83, FSHS.
[3] Ibid.
[4] Doug Scott's 5/25/72 Statement on S. 866 before the Senate Interior Committee,

Denver Conservation Library (DCL), Wilderness Society Records (WSR), Box 5, French Pete Box, Forest Service Memo, 12/54.
[5] DCL, WSR, French Pete Box, Decision of David Gibney, 3/25/69.
[6] Interview with Brock Evans, Washington, DC, 7/15/83, FSHS.
[7] Ibid.
[8] DCL, WSR, French Pete Box, Sunday Oregonian, 8/20/72.
[9] DCL, WSR, French Pete Box, Eugene Register Guard, 11/1/73.
[10] Ibid.
[11] Winninette Noyes, letter to John Hall of The Wilderness Society, 4/10/69, DCL, WSR, French Pete Box.
[12] Holway Jones, letter to Senator Robert Packwood, 4/27/73, DCL, WSR, French Pete Box.
[13] DCL, WSR, French Pete Box, Sunday Oregonian, 8/20/72.
[14] Oregon Daily Emerald, 11/16/72.
[15] DCL, WSR, French Pete Box, Eugene Register Guard, 4/23/75.
[16] Interview with Larry Williams, Washington, DC, 7/21/83, FSHS.
[17] Ibid.
[18] Scott interview, cited above.
[19] Ibid.
[20] Ibid.
[21] Ibid.
[22] Ibid.
[23] "Report to Accompany H.R. 3454," report No. 95-540, House, 95th Congress, lst session, 1977.
[24] Scott interview, cited above.
[25] DCL, WSR, Box 21, draft letter from Michael McCloskey, no date.
[26] Scott interview, cited above.
[27] George Davis, letter to George Marshall, 12/21/76, DCL, WSR, Endangered American Wilderness Box.
[28] Scott inteview, cited above.
[29] Cutler interview, cited above.
[30] Ibid.
[31] Scott interview, cited above.
[32] Cutler interview, cited above.
[33] Interview with Max Peterson, Washington, DC, 2/26/84, FSHS.
[34] Cutler interview, cited above.
[35] Scott interview, cited above.
[36] DCL, WSR, Endangered American Wilderness Box, "Oregonian", 5/7/77; Memoir from Doug Scott to key wilderness leaders, 5/9/77.
[37] Interview with John Hall, Washington, DC, 7/25/83, FSHS.
[38] Records of Rupert Cutler, United Press International, n.d.
[39] Cutler interview, cited above.
[40] Tim Mahoney, letter to the author, 5/11/84, FSHS.
[41] "RARE II: Final Environmental Statement," Washington, DC:USDA Forest Service, 1979, p. 17.
[42] Mahoney letter, cited above.
[43] Interview with Tim Mahoney, Washington, DC, 6/15/83, FSHS.
[44] Lloyd Meads, speech before the Western States Legislative Forestry Task Force, 8/1/79, Sierra Club (SC) records, FSHS.
[45] Interview with Andy Kerr, Eugene, OR, 3/11/85.
[46] "Forestry Industry Affairs Letter," 4/13/81.
[47] Interview with Tim Mahoney, Washington, DC, 4/3/85.
[48] Dennis M. Roth, The Wilderness Movement and the National Forests: 1964-1980 (Washington, DC: U.S. Department of Agriculture, Forest Service; 1984), p. 58.
[49] Mahoney interview, cited above.
[50] Ibid.; Doug Scott to John McComb, 6/26/80, Records of the Sierra Club, FSHS.
[51] Roth, The Wilderness Movement, p. 59.
[52] Ibid., p. 66.
[53] Dick Barnes, telephone interview, 4/23/85.
[54] John Seiberling to Richard L. Barnes, 6/11/81, House Public Lands Subcommittee.
[55] Interview with Andy Wiessner, Washington, DC, 11/20/84.
[56] Twin Falls, ID, "Times-News," 4/16/81.
[57] Wiessner interview, cited above.
[58] Interview with Scott Shotwell, Washington, DC, 12/4/84.
[59] "Lewiston [Idaho] Morning Tribune," 3/18/84.
[60] Mahoney interview, cited above.
[61] Cutler interview, cited above; "Public Land News," 4/30/81.
[62] Mahoney interview, cited above.
[63] Ibid.
[64] "Eugene Register Guard," 11/26/81.
[65] "Public Land News," 12/10/81.
[66] Statement of Peter D. Coppelman. . . on S. 543 and H.R. 1568, 6/28/ 83, The Wilderness Society (TWS) records, FSHS.
[67] "Steve Forrester's Northwest Letter," 4/30/84.
[68] Mahoney interview, cited above.
[69] Marc Leepson, "Protecting the Wilderness," Editorial Research Reports, vol. 11, no. 6, Aug. 17, 1984, p. 600.
[70] Alec Dubro, "Wilderness on Trial," "Outside," Feb./March 1982.
[71] Mahoney interview, cited above; Casper, WY, "Star-Tribune," 10/10/80.
[72] Ibid.
[73] "Missoulian," 5/10/81.
[74] "Billings Gazette," 4/12/80.
[75] Mahoney interview, cited above.
[76] Ibid.
[77] Ibid.
[78] "Wall St. Journal," 8/28/81; "Sierra Club National News," 8/31/81.
[79] "Billings Gazette," 9/10/81
[80] Mahoney interview, cited above.
[81] "Missoulian," 11/20/81, James Watt to Manuel Lujan, Jr., 11/19/81, SC records, FSHS.
[82] Mahoney interview, cited above.
[83] Tim Mahoney and John McComb to Doug Scott, 11/20/81, SC records, FSHS.
[84] "Denver Post," 1/30/82.
[85] Mahoney interview, cited above.
[86] Ibid.
[87] "Los Angeles Times," 2/24/82.
[88] Mahoney interview, cited above.
[89] Ibid.
[90] "Los Angeles Times," 9/24/82.
[91] "New York Times," 1/27/83; "Weekly Bulletin," 12/13/82.
[92] Russ Shay to members, 8/13/82, SC records, FSHS.
[93] Interview with Mark Reimers, Washington, DC 10/30/85.
[94] "East Linn County: A Case Study in the Effects of Wilderness," Oregon Women for Tim-

ber, 5/83, SC records, FSHS.
[95] Mahoney interview, cited above.
[96] Tim Mahoney to Joe Fontaine, et al., 8/18/81, SC records, FSHS.
[97] Mahoney interview, cited above.
[98] Shotwell interview.
[99] Mahoney interview, cited above.
[100] Interview with Tom Thompson, Washington, DC, 12/4/84.
[101] Ibid.
[102] "Statement of Peter D. Coppelman. . . on H.R. 3960, 1/25/84, TWS records, FSHS; interview with John Melcher, Washington, DC, 4/26/85.
[103] Ibid.
[104] Thompson interview, cited above.
[105] Interview with Peter Coppelman, Washington, DC, 1/18/85.
[106] Tim Mahoney et al. to John Seiberling, 12/12/83, SC records, FSHS.
[107] Kerr interview, cited above.
[108] Telephone interview with Tom Imeson, 4/8/85.
[109] Mahoney interview, cited above.
[110] Ibid.; interview with Peter Coppelman, Washington, DC, 9/20/84.
[111] Mahoney interview, cited above; "Oregonian," 4/2/84.
[112] Interview with Bill Brooks, Washington, DC, 12/13/84.
[113] Tim Mahoney to field representatives, 3/28/84, SC records, FSHS.
[114] "Congressional Quarterly," March 31, 1984, pp. 727-28.
[115] U.S. Senate Energy Committee Business Meeting, 4/11/84, Acme Reporting.
[116] Ibid.
[117] "Congressional Quarterly," March 31, 1984, pp. 727-28.
[118] Shotwell interview, cited above; interview with Max Peterson, Washington, DC, 9/15/84.
[119] Interview with Barbara Wise, Washington, DC, 1/28/85.

Chapter VII

[1] Robert T. Wazeka, "Organizing for Wilderness," Sierra Club Bulletin, 10/76, p. 50.
[2] Ibid., p. 50.
[3] Ibid., p. 49.
[4] Imeson interview, cited above.
[5] OWC/YCCIP application, n.d., Holly Jones' Files, FSHS.
[6] Jim Montieth to Sierra Club National Wilderness Committee, 2/15/77, Holly Jones' Files, FSHS.
[7] Kerr interview, cited above.
[8] Mahoney interview, cited above.
[9] Interview with Clif Merritt, Denver, CO, 3/15/85; Wazeka, "Organizing," p. 51.
[10] Wazeka, "Organizing," p. 50.
[11] Ibid., p. 51.
[12] Jim Montieth to OWC Executive Committee, 5/8/76, Holly Jones' Files, FSHS.
[13] Wazeka, "Organizing," p. 50.
[14] Jim Montieth to OWC Executive Committee, 5/8/76.
[15] Jim Montieth to Sierra Club National Wilderness Committee, 2/15/77.
[16] Interview with Jim Montieth, Washington, DC, 4/3/85.
[17] Kerr interview, cited above.
[18] Ibid.
[19] Ibid.
[20] Montieth interview, cited above.
[21] Richard Worthington to Forest Supervisors and Directors, 12/30/81, SC records, FSHS.
[22] Kerr interview, cited above; Imeson interview, cited above; "Oregonian," 7/24/83; "Wild Oregon," Fall 1984, p. 7.
[23] Kerr interview, cited above; Imeson interview, cited above; "Oregonian," 7/24/83; "Wild Oregon," Fall 1984, p. 7.
[24] Montieth interview, cited above.
[25] Interview with Arnold Ewing, Western Oregon Timber Association, Eugene, OR, 3/11/85.
[26] Imeson interview, cited above.
[27] Kerr interview, cited above.
[28] Ibid., Mahoney interview, cited above.
[29] Mahoney interview.
[30] Ibid.
[31] Interview with Greg Skillman, Eugene, OR, 3/12/85.
[32] Ibid.
[33] Ibid.
[34] Rick Battson to Steve Levy, 10/17/80, SC records, FSHS.
[35] Jim Montieth to Forest Wilderness Leaders, 2/13/81; telephone interview with Ron Eber, 3/18/85.
[36] Jim Montieth to Oregon Forest Leaders, 12/22/81, SC records, FSHS.
[37] Interview with Gerald Williams, Eugene, OR, 3/24/86.
[38] Interview with Steve Evered, former aide to Les AuCoin, Washington, DC, 3/27/85.
[39] Skillman interview, cited above.
[40] Evered interview, cited above; "Eugene Register-Guard," 12/17/82.
[41] Mahoney interview, cited above.
[42] Kerr interview, cited above.
[43] Mahoney interview, cited above; "Steve Forrester's Northwest Letter," 6/1./84.
[44] Interview with Rolf Anderson, Eugene, OR, 3/24/86.
[45] Imeson interview, cited above.
[46] Ibid.
[47] Jean Durning to Peter Coppelman, 11/3/83, Peter Coppelman's Files, FSHS.
[48] Imeson interview, cited above.
[49] Interview with Mike Kerrick, Eugene, OR, 3/24/86.
[50] Skillman interview, cited above.
[51] "East Linn County: A Case Study in the Effects of Wilderness," Oregon Women for Timber, 5/1/83, SC records, FSHS.
[52] Kerrick interview, cited above.
[53] Eber interview, cited above.
[54] Wendell Wood, president of ONRC, to the Oregon Sierra Club, 3/19/84, SC records, FSHS.
[55] Montieth interview, cited above.
[56] Interview with Jean Durning, Washington, DC, 3/27/85.
[57] Imeson interview, cited above.
[58] Interview with Holly Jones, Eugene, OR, 3/12/85.
[59] Skillman interview, cited above.
[60] Durning interview, cited above; "Oregonian," 4/2/84.
[61] Imeson interview, cited above.
[62] Interviews with Jack Wright and Dick Schwarzlender, Roseburg, OR, 3/25/86.
[63] Interview with Mike Anderson, Washington, DC, 3/27/85.
[64] Ewing interview, cited above.

[65] Jim Montieth to Governing Council Members, 4/12/83, SC records, FSHS.
[66] Kerr interview, cited above.
[67] Anderson interview, cited above.
[68] Skillman interview, cited above.
[69] Mahoney interview, cited above.
[70] Kerr interview, cited above.
[71] Jim Montieth to Tim Wapato et al., 12/2/83, SC records, FSHS.
[72] Imeson interview, cited above.
[73] Montieth interview, cited above.
[74] Eber interview, cited above.
[75] Mahoney interview, cited above.
[76] Evered interview, cited above; "Wild Oregon," Fall 1984, p. 22.
[77] Eber interview, cited above.
[78] Kerr interview, cited above.
[79] Mahoney interview, cited above.
[80] Eber interview, cited above.
[81] Mahoney interview, cited above.

Chapter VIII

[1] U.S. House of Representatives, Hearings Before the Subcommittee on Public Lands, 88 Cong., 2d sess., April 27-30, May 1, 1964, p. 1227.
[2] Principal Laws Relating to Forest Service Activities, 197.
[3] Ibid., p. 303.
[4] McCloskey, "Wilderness Act," p. 311; U.S. Senate, Committee on Interior and Insular Affairs, Hearings on S. 4028, 85 Cong., 2d sess., 1959, p. 257; Michael McCloskey, "What the Wilderness Act Means," in Action for Wilderness, edited by Elizabeth R. Gillette (San Francisco: Sierra Club, 1972), pp. 14-21.
[5] Hearings, p. 158.
[6] Joel Gottlieb, "The Preservation of Wilderness Values: The Politics and Administration of Conservation Policy," Ph.d. dissertation, University of California at Riverside, 1972, p. 184.
[7] Ibid., p. 190; Ronald Gibson Strickland, "Ten Years of Congressional Review Under the Wilderness Act of 1964: Wilderness Classification Through 'Affirmative Action,'" Ph.D. dissertation, Georgetown University, 1976, pp. 226-27.
[8] Roth, The Wilderness Movement, p.61.
[9] Mahoney interview, 4/3/85.
[10] Interview with Bruce Hamilton, San Francisco, CA, 3/4/85.
[11] Steven Lewis Yaffee, The Wisdom of the Spotted Owl; Policy Lessons for a New Century (Washington DC: Island Press, 1994), p. 74.